# Oil and Gas Wells

*Edited by Sid-Ali Ouadfeul*
*and Leila Aliouane*

Published in London, United Kingdom

IntechOpen

*Supporting open minds since 2005*

Oil and Gas Wells
http://dx.doi.org/10.5772/intechopen.78185
Edited by Sid-Ali Ouadfeul and Leila Aliouane

Contributors
Eman Mansour, Emilson Pereira Leite, Steve Brenner, Adesina Fadairo, Abiodun Adeyemi, Temitope Ogunkunle, Ayotomiwa Evbogame, Sid-Ali Ouadfeul, Leila Aliouane, Emmanuel Imarhiagbe, Nosa Omoregbe Obayagbona, M. Karimi



First published in London, United Kingdom, 2020 by IntechOpen
IntechOpen is the global imprint of INTECHOPEN LIMITED, registered in England and Wales, registration number: 11086078, 7th floor, 10 Lower Thames Street, London, EC3R 6AF, United Kingdom
Printed in Croatia

British Library Cataloguing-in-Publication Data
A catalogue record for this book is available from the British Library

Additional hard and PDF copies can be obtained from orders@intechopen.com

Oil and Gas Wells
Edited by Sid-Ali Ouadfeul and Leila Aliouane
p. cm.
Print ISBN 978-1-78984-630-0
Online ISBN 978-1-78984-631-7
eBook (PDF) ISBN 978-1-83880-137-3

For EU product safety concerns:
IN TECH d.o.o., Prolaz Marije Krucifikse Kozulić 3, 51000 Rijeka, Croatia, info@intechopen.com or visit our website at intechopen.com.

# Meet the editors

Dr Sid-Ali Ouadfeul is currently an associate professor of geophysics at Khemis Miliana University-Algeria. He received an Engineer diploma from the University of Boumerdes, Algeria, a Magister and PhD in geophysics from the University of Sciences and Technology Houari Boumedienne in Algeria. One of the topics of Dr Ouadfeul's research is oil and gas; he has published many papers related to oil and gas exploration and reservoir characterization.

Dr Leila ALIOUANE is a teacher-researcher in the Faculty of Hydrocarbon and Chemistry - University M'hamed Bougara of Boumerdès (UMBB). Her main area of interest is oil and gas reservoir characterization from seismic and well-log data using advanced signal processing techniques.

# Contents

# Preface

Oil and gas are the most useful forms of fossil energy and they represent more than 80% of the world's energy production. With an increasing demand on this energy in the last few decades due to rapid development of the world's industries, scientists should adapt the latest technology to this demand. The aim of this book is to present some advances in different aspects of oil and gas technology. This book is edited by Dr Sid-Ali Ouadfeul, an associate professor in geophysics at Khemis Miliana University. Two chapters of the book are dedicated to scientific research in the domain of reservoir engineering and characterization. Four chapters are dedicated to the field of well drilling and performance and another chapter is related to oil and transport.

**Dr. Sid-Ali Ouadfeul**
Algerian Petroleum Institute,
Algeria

**Dr. Leila Aliouane**
University of Boumerdes,
Algeria

Chapter 1

# Introductory Chapter: Oil and Gas Wells - Advances and New Challenges

*Sid-Ali Ouadfeul and Leila ALiouane*

## 1. Introduction

Oil and gas are the most useful fossil energy; they are presenting more than 80% of the world energy production (see **Figure 1**), with the increasing demand of these energy in the last decades due to rapid development of the world industries. Exploration, production, transport, refining, and commercialization of oil and gas require new methods and procedures to satisfy the needs of the different industrial sectors and world population in terms of fuel energy.

A study by Hull [2] (a Halliburton Consulting) shows that the production of oil and gas in the world is under the economic limit since 2010, and it continues to decrease until 2030; the peak of production was in 1968 (see **Figure 2**). Another aspect showed in this report that when talking about mature fields is the concept of economic limit. The fact that we only recover on average 35% of the oil in place globally is not a function of technology or know-how, but rather it is dictated by what is economic to extract. The challenge for oil companies and researchers, therefore, is finding and applying technology and know-how that allows us to extract the resources at a cost that achieves the economic threshold [2].

For example, in the oil and gas domain, we can distinguish two kinds of oil and gas types which are conventional and unconventional; they have the same chemical characteristics and components; the only difference between them is in their way of extraction, since the conventional oil and gas are small quantities easy to develop with low cost; however the unconventional hydrocarbons are huge quantities requiring

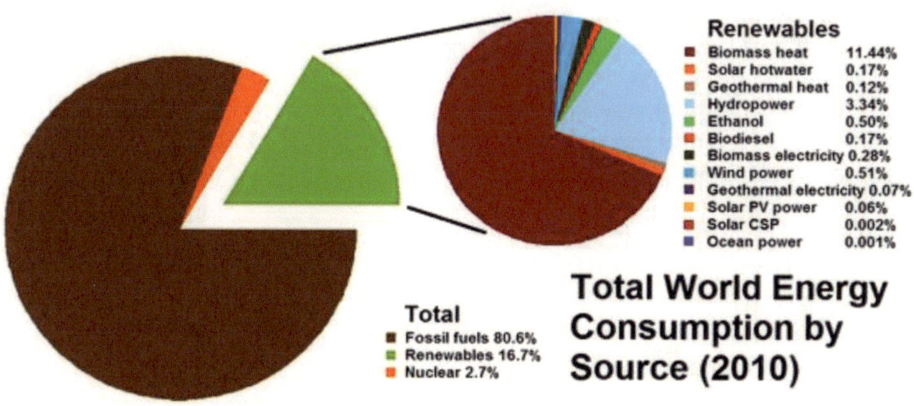


**Figure 1.**
*World energy development [1].*

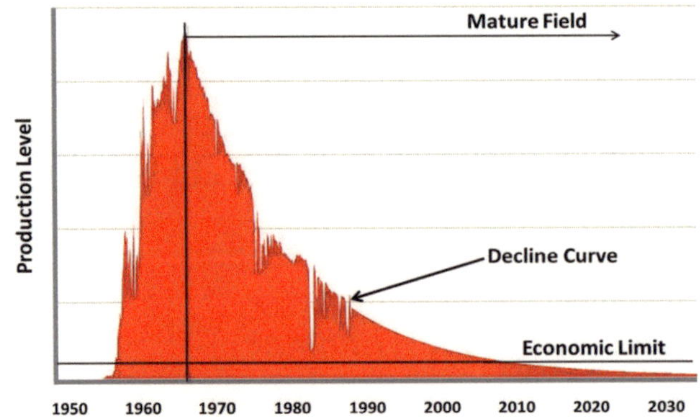


**Figure 2.**
*Oil and gas production curve (Hull, 2012).*

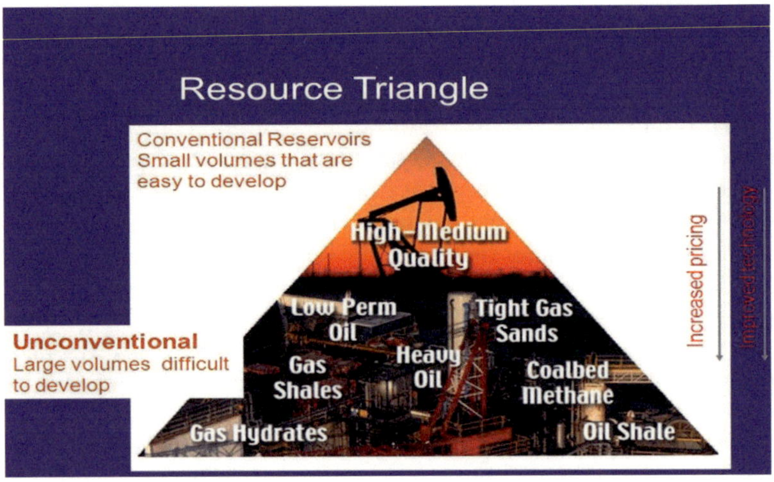


**Figure 3.**
*Oil and gas resources triangle (from [4]).*

advanced technology to develop them with increasing price [3, 5] **Figure 3**. Shows the different kinds of conventional and unconventional oil and gas resources.

Each phase in the oil and gas field requires continuous scientific research to improve the methodologies and procedures used in different stages inside this phase. The aim of this book is to show some advances in different topics of the oil and gas field technology; two chapters of the book are dedicated to the scientific research in the domain of reservoir engineering and characterization, four other chapters are dedicated to the field well drilling and performance, while another chapter is related to oil and transport.

## Author details

Sid-Ali Ouadfeul[1*] and Leila ALiouane[2]

1 University of Khemis Miliana, Algeria

2 LABOPHYT, Geophysics Department, Faculty of Hydrocarbons and Chemistry, University of Boumerdes, Algeria

*Address all correspondence to: souadfeul@ymail.com

## References

[1] REN21–Renewable Energy Policy Network for the 21st Century Renewables. Global Status Report, page 21; 2012. Archived December 15, 2012, at the Wayback Machine

[2] Hull R. What is a mature field? Haliburton Report; 2012

[3] Ouadfeul S, Aliouane L. Shale gas reservoirs characterization using neural network. Energy Procedia. 2014;**59**: 16-21. DOI: 10.1016/j.egypro.2014.10.343

[4] Holdith S. Tight gas sands. Journal of Petroleum Technology. 2006;**58**(6): 84-90. DOI: 10.2118/103356-JPT

[5] Aliouane L, Ouadfeul S. Sweet spots discrimination in shale gas reservoirs using seismic and well-logs data. A case study from the worth basin in the barnett shale. Energy Procedia. 2014;**59**:22-27. DOI: 10.1016/j.egypro.2014.10.344

Chapter 2

# Porosity Prediction of a Carbonate Reservoir in Campos Basin Based on the Integration of Seismic Attributes and Well Log Data

*Roberta Tomi Mori and Emilson Pereira Leite*

## Abstract

We have calculated and interpreted a 3D porosity model of a reservoir through the integration of 3D seismic data with geophysical well logs using an artificial neural network (ANN). The reservoir is composed of Albian carbonates. In the first main stage of the study, horizons were traced by following continuous seismic events on seismic sections, along depths between top and base of the reservoir. In the second main stage, predictions of reservoir porosity values were obtained, as well as a 3D model, through the designed ANN. The estimated porosity values range from 5 to 30%. The correlation coefficient and the error of the estimated values with respect to the actual values extracted along the wells are equal to 0.90 and 2.86%, respectively. Porosity values increase from southwest to the northeast portion, and lower values are found at depths related to the traced horizons. Although isolated peaks of maximum porosity are observed, spatial patterns depicted in the model are associated with geological features such as different porosity types and cementation degree.

**Keywords:** porosity, artificial neural networks, carbonate reservoir

## 1. Introduction

Reservoir characterization has become increasingly important to hydrocarbon exploration. Accurate characterization reduces the risk of drilling a dry well, as well as exploration and development costs. For this reason, different types of data are used, such as geophysical well logs, and seismic, petrophysical, in addition to geological models, in order to predict reservoir properties such as porosity, lithology, and fluid saturation [1]. In particular, the integration of well logs with seismic data is important in order to obtain some models with better vertical and horizontal resolutions, since well logs have a very restricted area and a better vertical resolution when compared to seismic data; however, seismic data presents a better horizontal resolution and covers a larger area. Integrated quantitative interpretation is used to estimate reservoir properties, obtained through seismic amplitudes and seismic attributes [2–4].

A seismic attribute is any direct or indirect information obtained from the seismic data through mathematical calculation and/or logical reasoning. Depending on how it is derived, an attribute may help the interpreter to delineate geologic structures, map geologic features, estimate physical properties, etc. [4]. Because each independent attribute provides a particular view of the seismic data, the use of a single attribute leads to a high uncertainty in interpretation [5]. Therefore, the application of multiattribute analysis has grown during the last few decades [2, 6–8]. Multiattribute analysis employs a combination of various seismic attributes through mathematical modeling in order to increase the accuracy in the prediction of a particular property.

In this context, artificial neural networks (ANN) are tools to perform a multiattribute analysis. They allow us to establish a quantitative relationship between the well log data and the seismic data, such that it can be used to predict a physical property in positions where there are no well log data. In general, an ANN is composed of an input layer, an output layer, and one or more intermediate layers that are hidden. In the input layer, there are neurons that represent the input dataset. In the hidden layers, the neurons adjust the input data to the target well log values in the output layer, through an iterative calculation of weights. These weights define the model that is used for prediction of unknown values [3, 7, 9, 10].

In this chapter, we have predicted porosity values of a carbonate reservoir located on Campos Basin through an ANN method, applied to the integration of well log and 3D seismic data. This process provided a 3D numeric volume of porosity of the entire reservoir. We have interpreted the spatial distribution of the porosity values according to geological information obtained from the literature and from descriptions of core samples.

## 2. Geology of the study area

The Campos Basin is located offshore of the southeastern portion of the Brazilian continental margin (**Figure 1**). It encompasses an area of ~120,000 km$^2$ and the maximum water depth is ~3400 m. It is limited by the Vitoria Arc on the north, the Cabo Frio Arc on the south, the boundary of the salt diapir region at water depths of ~2200 m on the east, and the updip limits of the turbidites to the west. Campos Fault divides the deepest part (east portion), where cretaceous

**Figure 1.**
*Some of the offshore sedimentary basins of Brazil. Campos Basin is highlighted in red (source accessed in April 2018: http://wdetail.asp?img_id=4775&a_id=117349).*

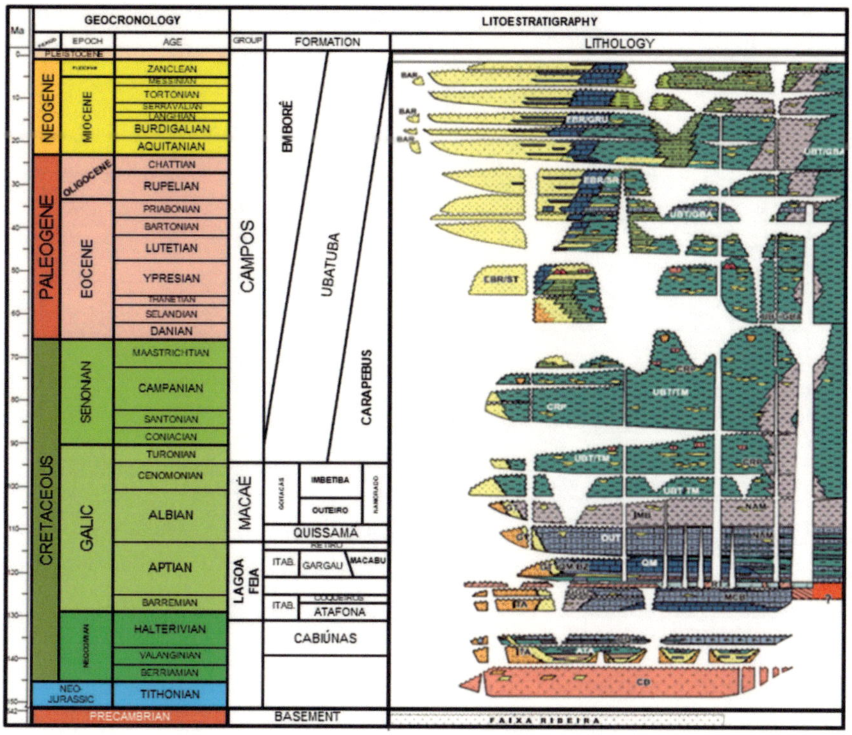


**Figure 2.**
*Simplified stratigraphic column of Campos Basin (adapted from: [12]).*

sediments were deposited, from the west portion where the sediments were deposited on top of the basement [11].

The studied reservoir is inserted into the Quissamã formation, which was formed during the final stage of the lower Cretaceous (Albian). This formation is composed of byoncolitic, oolitic, and peloid dolomites, poorly sorted sandstones, polymictic conglomerate associated to calcilutites and marls, and pelitic sediments dominated by marl ([12]; **Figure 2**). Quissamã formation is contained in the Macae Group, which has a shallow carbonate platform dominated by thick shoal carbonates with ooliths and oncolites. The reservoir consists mainly of oncolytic calcarenites and calcirudites, distributed in bars forming a NE trend. It essentially contains microporosity varying between 15 and 30% and has low permeabilities, but the presence of a fracture system contributes to the increasing of permeability in some regions [13].

## 3. Materials and methods

We have used profiles of neutron porosity, bulk density, gamma ray and sonic travel time from well logs, and a 3D seismic data. P-wave velocity was calculated from sonic logs. HRS (Hampson Russell Software[1]—CGG Veritas) was employed to perform time-depth conversions through seismic-well ties. These tie processes were applied within a time window analysis ranging from the top markers to the base markers. The overall correlation obtained was 0.65, which is not uncommon for this type of application (e.g., [14]).

We then identified continuous reflection events on the seismic sections and defined seismic horizons from the interpolation of picked time/distance pairs. This

[1]http://www.cgg.com/hampson-russell.aspx?cid=3609.

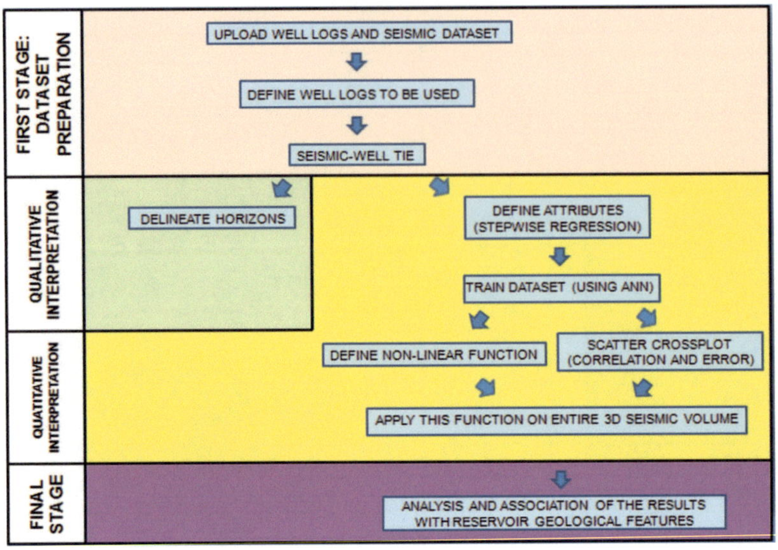


**Figure 3.**
*Flowchart that describes the process applied to obtain our 3D porosity model of the reservoir.*

is a qualitative interpretation which does not take into account the seismic amplitude but only the P-wave travel time.

After horizon tracing, we designed and trained an artificial neural network (ANN) to construct a porosity model of the studied reservoir. In the final interpretation stage, we analyzed the results and made the appropriated associations with the geological features found in the literature. These interpretations are also consistent with descriptions of core samples and petrophysical analysis. **Figure 3** shows a simplified scheme of the overall process applied in this work.

### 3.1 Multiattribute analysis

The general goal of a multiattribute analysis is to find a mathematical relationship between target reservoir properties and seismic attributes. Assuming that seismic-well ties were already conducted, two main stages are necessary to perform a multiattribute analysis (e.g., after this relationship is established along the wells, it is applied to populate the 3D seismic space with the chosen reservoir property (e.g., [9]):

1. To train the seismic attributes along the wells, so that they are mapped onto the desired property space. The most appropriated attributes are defined during this training.

2. To make predictions of the desired property for the entire seismic volume, using the mathematical relationship found on the first stage.

### 3.2 Probabilistic neural network

In this chapter, we performed a multiattribute analysis using a probabilistic neural network (PNN). This type of ANN has been described, for instance, by Specht [15] and by Masters [16, 17]. It basically consists in an interpolation scheme that uses the architecture of an ANN in its implementation. For example, consider $n$ training samples where each sample is composed of two input attributes, $A_1$ and $A_2$, and one output property value $L$ each, which is the measured target log value. In matrix form, this training set is a $n \times 3$ matrix whose entries are $\{A_{1i}, A_{2i}, L_i\}$ where $i = 1, \dots n$.

A PNN assumes that each new output value can be described as a linear combination of the input values in the training set. For a new input sample, $x = \{A_{1j}, A_{2j}\}$; an output value $L'(x)$ is estimated as

$$L'(x) = \frac{\sum_{i=1}^{n} L_i \, exp\left(-D(x,x_i)\right)}{\sum_{i=1}^{n} exp\left(-D(x,x_i)\right)}, \tag{1}$$

where

$$D(x,x_i) = \sum_{i=1}^{2} \left[\frac{x_j - x_{ij}}{\sigma_j}\right]^2. \tag{2}$$

$D(x,x_i)$ is the distance between the input data and each training point $x_i$ (**Figures 4** and **5**). This distance is measured in a multidimensional attribute space and normalized by the quantity $\sigma_j$ [2].

The PNN training consists in determining the best group of smoothing parameters $\sigma_j$. The criterion used to determine these parameters is a minimum validation error [2].

The validation result for the $m$th target sample is defined as

$$L'(x_m) = \frac{\sum_{i \neq m} L_i \, exp\left(-D(x_m,x_i)\right)}{\sum_{i \neq m}^{n} exp\left(-D(x_m,x_i)\right)}. \tag{3}$$

This is the predicted value of the $m$th target sample when this sample is left out of the training dataset. In other words, this means that the dataset is trained without that sample. By repeating this process for each training sample, total prediction error can be defined as

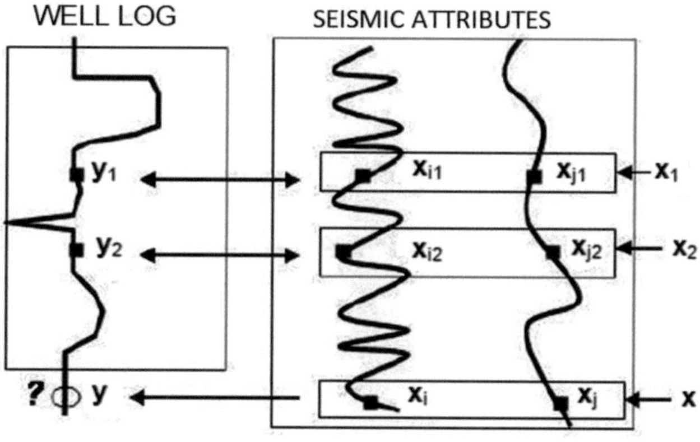


**Figure 4.**
*Representation of the positions at which the input vectors ($x_1$ and $x_2$) are associated to known target values of an output vector ($y_1$ and $y_2$) and a position where an input vector (x) is associated to an unknown target value of an output vector (extracted from [5]).*

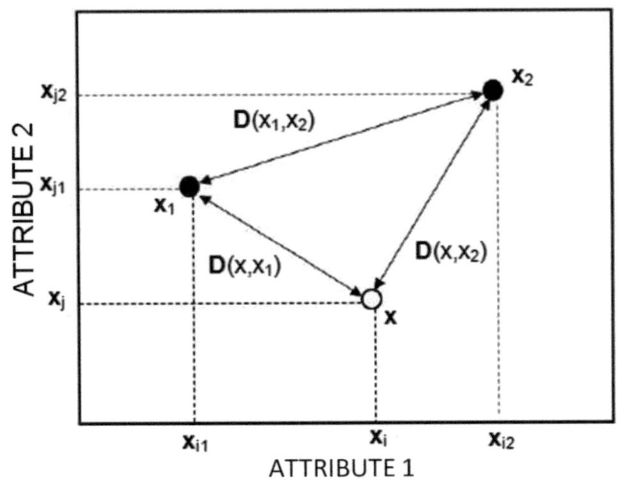


**Figure 5.**
*Schematic graph of vectors $x_1$, $x_2$, and x, relative to* **Figure 4** *where the coordinate axes represent the attribute amplitudes instead of Cartesian distances (modified from [5]).*

$$E(\sigma_1, \sigma_2, \sigma_3) = \sum_{i=1}^{N} \left(L_i - L'_i\right)^2. \tag{4}$$

## 4. Results and discussion

We have defined four horizons that were delineated by tracking continuous seismic events (**Figure 6**). The deepest one (**Figure 6a**) represents the reservoir base, while the shallowest (**Figure 6d**) represents the reservoir top. Between these horizons, there are more two horizons, termed intermediate horizons (**Figure 6c** and **d**). Drill core data show carbonate textures such as packstones and wackestones at depths near these horizons. These textures are composed by rock matrices rich in carbonate, which explains the decrease of porosity in these regions.

The PNN method was applied in the interval between the reservoir base and top horizons. After the training process, we have obtained a seismic attribute list that yields the highest correlation between predicted porosity and actual porosity at the wells and the lowest prediction error, through a stepwise regression. The attribute list and the total training error are presented in **Table 1**.

Then, using this attribute list, we applied the training process through the PNN using 12–19 attributes. Each training result was analyzed based on a correlation coefficient and a training error (**Figure 7b** and **c**). Based on the correlation and error values, 17–19 attributes yield the best predictions. Therefore, we chose the minimum number of attributes that produces the highest correlation and the lowest error. The prediction power of the trained PNN can be observed in the scatter plot of **Figure 7a**. The correlation coefficient is ~0.9 and the fitting error is ~2.86.

The trained PNN was applied to predict porosity values in the entire 3D seismic volume. **Figures 8a**, **b and 9a–d** show the porosity distribution along vertical and time slices extracted from the 3D porosity model. The spatial distribution of this model can also be observed in **Figure 9e** and **f**.

This model shows high heterogeneity in the distribution of the predicted porosity values, which vary from 5 to 30%, with an average of around 19%. These values

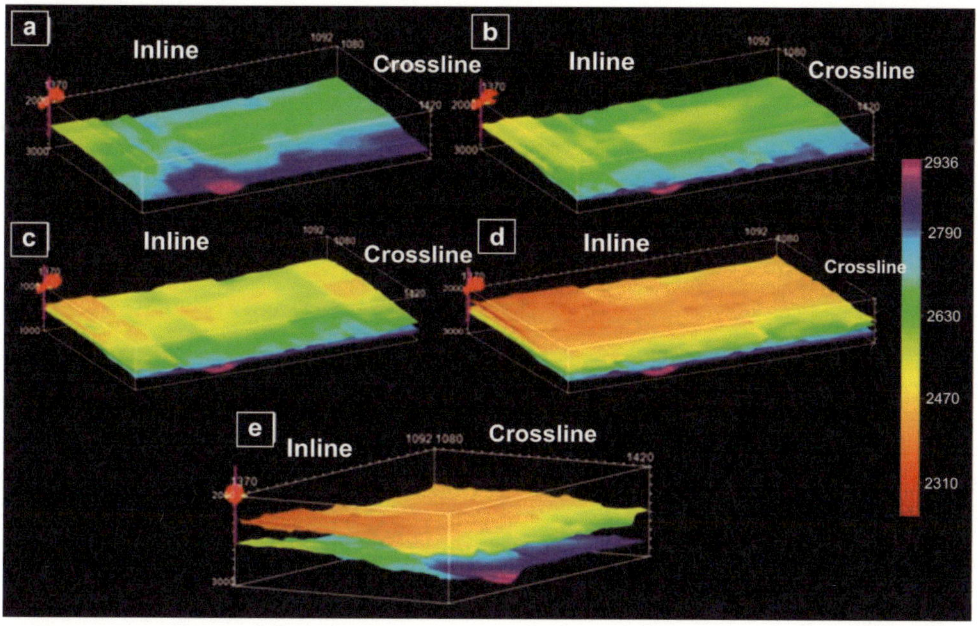


**Figure 6.**
*(a) Deepest horizons representing reservoir base, (b) deepest intermediate horizon, (c) the shallowest intermediate horizon, (d) the shallowest horizon representing reservoir top, and (e) reservoir top and base horizons.*

| | Target log | Attribute | Training error |
|---|---|---|---|
| 1 | Porosity | Dominant frequency | 6.066539 |
| 2 | Porosity | Amplitude weighted phase | 5.847074 |
| 3 | Porosity | Average frequency | 5.770756 |
| 4 | Porosity | Integrated absolute amplitude | 5.677001 |
| 5 | Porosity | Apparent polarity | 5.634725 |
| 6 | Porosity | Instantaneous frequency | 5.595683 |
| 7 | Porosity | Amplitude weighted cosine phase | 5.577585 |
| 8 | Porosity | Amplitude envelope | 5.552585 |
| 9 | Porosity | Amplitude weighted frequency | 5.512237 |
| 10 | Porosity | Quadrature trace | 5.496368 |
| 11 | Porosity | Instantaneous phase | 5.484175 |
| 12 | Porosity | Cosine instantaneous phase | 5.454468 |
| 13 | Porosity | Second derivative instantaneous amplitude | 5.441876 |
| 14 | Porosity | Filter 5/10–15/20 | 5.432924 |
| 15 | Porosity | Filter 15/20–25/30 | 5.423227 |
| 16 | Porosity | Integrate | 5.414948 |
| 17 | Porosity | Filter 25/30–35/40 | 5.409717 |
| 18 | Porosity | Second derivative | 5.407269 |
| 19 | Porosity | Derivative | 5.407243 |

**Table 1.**
*Attribute list obtained after training with PNN and stepwise regression.*

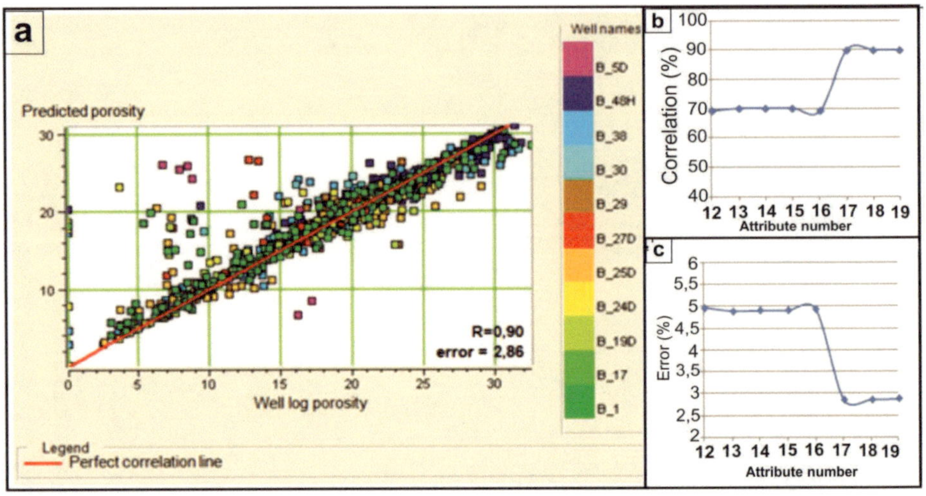


**Figure 7.**
*(a) Correlation between predicted porosity and well log porosity and the errors related to this correlation, obtained by probabilistic ANN. (b) Correlation between predicted porosity and actual porosity of PNN training using 12–19 attributes. (c) PNN training error.*

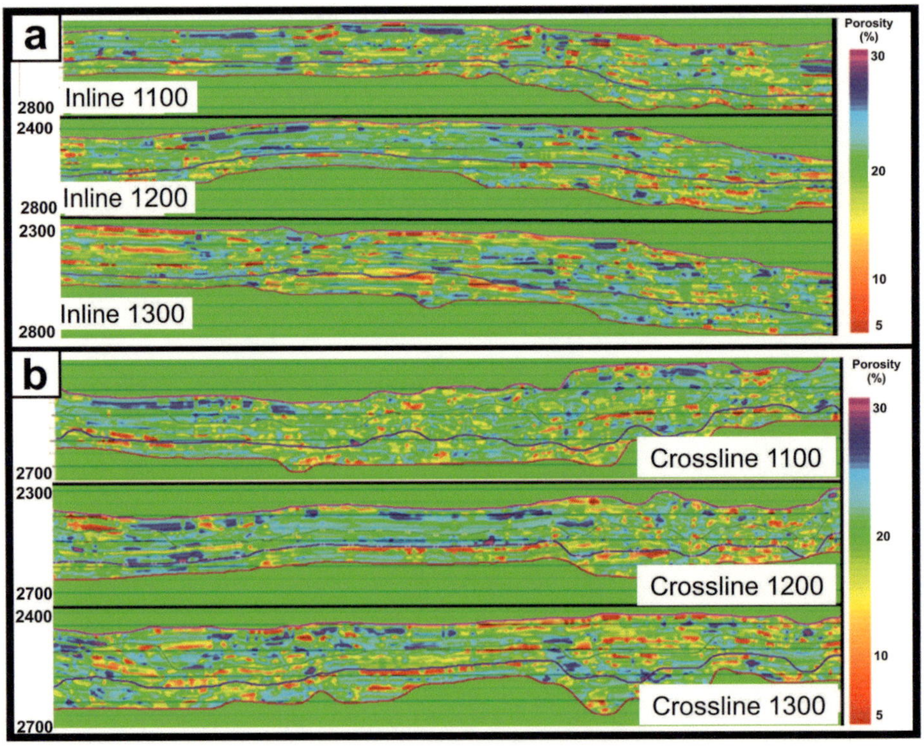


**Figure 8.**
*(a) Predicted porosity obtained by a PNN along inlines 1100, 1200, and 1300. (b) Predicted porosity obtained by a PNN along crosslines 1100, 1200, and 1300.*

increase from south to north and from west to east, with the highest values occurring on the northeast and on the central portion. Besides the horizontal porosity variation, we can also observe a significant vertical variation, in which the highest values are concentrated on the intermediate depths of reservoir, and there is also a decrease

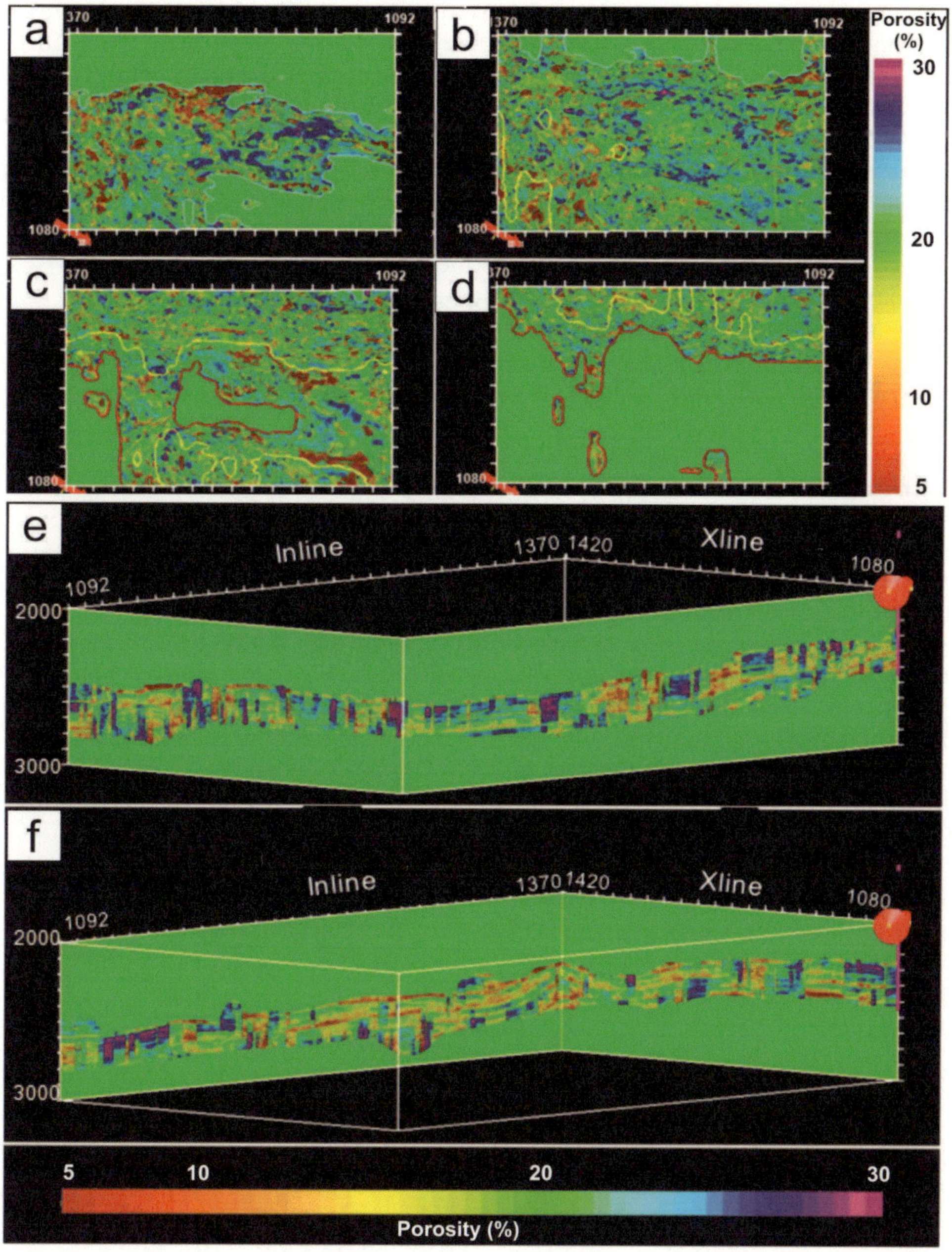


**Figure 9.**
*(a) A plan view of predicted porosity obtained by probabilistic ANN in 2400 ms, with north pointing to the bottom right corner of the image. (b) Predicted porosity in 2500 ms. (c) Predicted porosity in 2600 ms. (d) Predicted porosity in 2700 ms. (e) 3D volume of predicted porosity obtained by probabilistic ANN, on inline 1092 and crossline 1080. (f) 3D volume of predicted porosity on inline 1370 and crossline 1420.*

on these values on depths corresponding to the interpreted seismic horizons. Based on this information, we could identify three different groups of porosity: (1) those with low values (5–10%) and concentrated on the southwest portion and near seismic horizons; (2) those with intermediate values (10–22%) and dispersed in the entire seismic volume, however concentrating more on the northeast and on the central portion; and (3) those with high values (22–30%) concentrated in some specific regions, mostly on the northeast and on the central portion.

Melani [13] explains the difference between the porosity values on the northeast portion and on the southwest portion as the result of different porosity types. On the northeast, the porosity is the original macroporosity of the rocks, while in the southwest it is essentially microporosity. Besides this noticeable division, according to petrophysical and drill core data, porosity values vary due to rock texture variation of reservoir carbonates. Grainstones are related to higher porosity values, while packstones and wackestones are related to lower values, because of the difference in the degree of cementation.

Furthermore, the lower porosity values near the seismic horizons can be related to a higher cementation degree in these regions, where rock textures vary from packstone to wackestone. The variation of cementation degree occurred because of different energy availability in the depositional environment, which corresponds to drowning and shallowing cycles on these regions.

## 5. Conclusions

We have created a 3D porosity model of a carbonate reservoir in the Campos Basin through the application of a PNN that integrates well logs and seismic data. The overall correlation between predicted and actual porosity values is ~0.90, while the training error is ~2.86. This model presents high spatial heterogeneity. In general, porosity values increase from southwest to northeast, and the highest concentration is located on the northeast and central areas. This is explained by different porosity types in the reservoir, where the higher values are concentrated in regions where the original macroporosity was preserved and the lower values are concentrated where the porosity is classified as microporosity. While the shallowest horizon is related to the reservoir top and the deepest is related to the reservoir bottom, the two intermediate horizons are related to discontinuities with different cementation degrees. These horizons are associated to different rock textures caused by energy availability variation in the depositional environment. Our interpretations are based on petrophysical and drill core data available in the published literature.

## Acknowledgements

The authors thank the Institute of Geosciences at the University of Campinas for providing all resources to the development of this work. This study was financed in part by the Coordenação de Aperfeiçoamento de Pessoal de Nível Superior—Brasil (CAPES)—Finance Code 001, for which Roberta Tomi Mori is grateful for her Masters scholarship. Emilson Pereira Leite thanks Conselho Nacional de Desenvolvimento Científico e Tecnológico (CNPq) for his Productivity Grant #301593/2012-4.

## Author details

Roberta Tomi Mori and Emilson Pereira Leite*
Department of Geology and Natural Resources, Institute of Geosciences, University of Campinas, Campinas, SP, Brazil

*Address all correspondence to: emilson@ige.unicamp.br

## References


[1] Mohaghegh S, Arefi R, Ameri S, Aminiand K, Nutter R. Petroleum reservoir characterization with the aid of artificial neural networks. Journal of Petroleum Science and Engineering. 1996;**16**:263-274

[2] Hampson D, Schuelke J, Quirein J. Use of multi-attribute transforms to predict log properties from seismic data. Geophysics. 2001;**66**:220-236

[3] Leite EP, Vidal AC. 3D porosity prediction from seismic inversion and neural networks. Computers & Geosciences. 2011;**37**:1174-1180

[4] Taner MT. Seismic attributes. CSEG Recorder. 2001;**26**(7):49-56

[5] Russell BH. The application of multivariate statistics and neural networks to the prediction of reservoir parameters using seismic attributes [PhD thesis]. Canada: University of Calgary; 2004. 392p

[6] Brown AR. Understanding seismic attributes. Society of Exploration Geophysicists. 2001;**66**(1):47

[7] Khoshdel H, Riahi MA. Multi attribute transform and neural network in porosity estimation of an offshore oil field—A case study. Journal of Petroleum Science and Engennering. 2011;**78**:740-747

[8] Taner MT, Walls J, Taylor G. Seismic attributes, their use in petrophysical classification. In: EAGE 63° Conference & Technical Exhibition—Amsterdam, The Netherlands; 2001

[9] Schultz PS, Ronen S, Hattori M, Corbett C. Seismic guided estimation of log properties, parts 1, 2, e 3. The Leading Edge. 1994;**13**:305-310, 674-678, and 770-776

[10] Valenti JCAF. Porosity prediction from seismic data using multiattribute transformations, N sand, Auger Field, Gulf of Mexico [Master thesis]. United States: The Pennsylvania State University; 2009. 84p

[11] Guardado LR, Gamboa LAP, Lucchesi CF. Petroleum geology of the Campos Basin: A model for producing Atlantic type basin. In: Edwards JD, Santogrossi PA, editors. United States: Divergent/Passive Margin Basins. Vol. 48. AAPG Memoir; 1989. pp. 3-79

[12] Winter WR, Jahnert RJ, França AB. Bacia de Campos. Boletim de Geociências da Petrobras, Rio de Janeiro. 2007;**15**:511-529

[13] Melani LH. Caracterização petrofísica de reservatório carbonático [Dissertação de Mestrado]. Campinas: Faculdade de Engenharia Mecânica e Instituto de Geociências, Universidade Estadual de Campinas; 2015. 80p

[14] Sancevero SS, Remacre AZ, Portugal RS. O papel da inversão para a impedância acústica no processo de caracterizacão sísmica de reservatórios. Revista Brasileira de Geofísica. 2006;**24**(4):495-512

[15] Specht DF. Probabilistic neural networks. Neural Networks. 1990;**3**(1):109-118

[16] Masters T. Practical Neural Network Recipes in C++. United States: Academic Press, Inc; 1993

[17] Masters T. Advanced Algorithms for Neural Networks. United States: John Wiley & Sons, Inc; 1995

Chapter 3

# Environmental Evaluation and Biodegradability of Drilling Waste: A Case Study of Drill Cuttings from Ologbo Oilfield Wells at Edo State, Nigeria

*Emmanuel Esosa Imarhiagbe and Nosa Omoregbe Obayagbona*


## Abstract

Oil-laden drill cutting wastes have remained a serious environmental menace to well engineers and oil prospecting companies, due to unacceptability of oil-based muds to the environment as proscribed by the environmental guidelines. The problem of oil-containing drill cuttings can be better appreciated when viewed along the line that in Nigeria, about 3,900 billion barrel of drill cuttings are produced in a typical four thousand and fifty-four meter on shore drilling operation. Guidelines and standards of the regulatory authority in Nigeria, the Department of Petroleum Resources, forbid the discharge of drill cuttings into the environment without first ascertaining the nil or minimum impacts via carrying out Environmental Impact Assessment and Environment Evaluation Report studies. Biodegradation is the natural process whereby micro-organisms use up such substances as energy source, which are broken down into constituents such as fatty acids, carbon dioxide, and water. The biodegradation of oil pollutants is not a new concept; however, it is new as an increasingly effective and potentially inexpensive clean-up technology. Its potential contribution as countermeasure biotechnology for decontamination of oil-polluted ecosystems is enormous. Oil exploration industries should adopt biodegradation treatment procedures of their generated wastes before discharge into receiving environment.


**Keywords:** drill cuttings, biodegradation, guidelines, microorganisms, environment

## 1. Introduction

Drill cuttings are pieces of rocks that are generated when holes are drilled through the earth's crust to reach the oil and gas reservoir. Depending on the type of rock formation being drilled and the drilling rig being employed for the process, these small pieces of rocks can vary in size and texture, with particle size ranging from sand to gravel.

The environmental discharge of oil-laden drill cuttings have remained an unabated challenge to environmentalists and well engineers. This unabated challenge is majorly due to the presence of banned oil-based mud on the resultant drill cutting which cannot be discharged directly into the receiving environment without pretreatment to meet the regulatory standards contained in the environmental guidelines for petroleum industries in Nigeria DPR [1] and other standards in the world.

Rotary drilling rigs are used throughout the world's oil regions, and its greatest advantage over other methods, such as the cable-tool drilling, is that the well bore is kept full of liquid (drill mud) during drilling. The composition of drill cuttings also includes a suspension of solids such as clays, barite, etc. in liquids (water or oil) with chemical additives to modify their properties [2]. According to DPR [1] and Odokuma and Ikpe [2], the chemical additives include bactericides, lubricants, thinner, dispersant and pH control additive.

According to Soegianto et al. [3], water-based muds are by far the most commonly used drill muds, both in onshore and offshore drilling operations. Water-based muds are widely used in shallow wells and often in shallower portions of deeper wells. Depending on the depth and diameter of the well, the use of water-based muds generate about 1400–2800 billion barrels (bbl) wastes, of that amount are drill cuttings [3]. Water-based muds are very economical and easy to dispose because of their biodegradability and low toxicity quality. Oil-based muds were developed and refined to overcome the limitation of water-based muds applications [3–5]. Oil-based muds have been observed to possess technological advantage in producing lower waste volume over water-based muds especially when drilling for oil and gas in adverse environmental situations such as including high temperature, hydratable shales, high angle and extended-reach well, high density mud and drill through to salt [3].

The problem of oil-containing drill cuttings can be better appreciated when viewed along the line that in Nigeria, about 3900 billion barrel (bbl) of drill cuttings amounting to 1482 million tones are produced in a typical 4054 m onshore drilling operation [6]. An average offshore well on the other hand generates approximately 1100 million tones of drill cuttings which, prior to being restricted by legislation, contained roughly 15% by weight of oil [6].

Generally, it is known that the accumulation of drill cuttings coated with drilling mud can have localised effects on the seabed and soil-dwelling microorganisms, due to its heavily laden suspended matter, oil and grease and heavy-metal concentration [5, 7–9]. According to [10] petroleum exploration in Nigeria began as far back as 1908, and in 1956, the first commercial oil discovery was made at Oloibiri in the onshore Niger Delta, and the production of the crude oil commenced in 1958. Seven years after, Chevron discovered the Okan field, the first commercial offshore field in 1963 [11]. Since that time, petroleum has played a progressively prominent role in the social and economic development of Nigeria. It is therefore not surprising that today, petroleum resources account for about 99% of the national economy.

The geophysicists and geologists have shown that the Niger Delta basin has spectacularly maintained a thick sedimentary apron and salient petroleum geological feature. This feature is favourable for petroleum generation, expulsion and trapping from the onshore through the continental shelf and to the deepwater terrains [12, 13]. According to Ekweozor and Daukoru [13], among the sedimentary basins in Nigeria, aggressive exploration has been concentrated in the Niger Delta basins, which is to date the most prolific and economic sedimentary basin considering the extent of petroleum accumulations, exploration as well as the spatial distribution of the resources on the onshore, continental shelf through deepwater terrains.

Ologbo community is one of the oil-producing communities in Edo State. It shares part of its boundaries with the Koko community in the Delta state. Thus, the inhabitants are both the Edo- and Urhobo-speaking people living mutually alongside with other ethnic groups such as Igbos, Ijaws and Itsekiris. Currently, Ologbo community houses flow stations and oil and gas wells belonging to the Nigerian Petroleum Development Corporation and Pan Ocean Petroleum Corporation. The people of this area are predominantly farmers and traders. The major crops cultivated are yam, maize and cassava.

Several researches, which include Odokuma and Okpokwasili [12, 14, 15], have reported that increased exploration and production activities as well as improper waste disposal practices in the locations of boreholes had continued to encourage the widespread contamination of the receiving environments which include aquatic and terrestrial ecological systems. The Niger Delta region in Nigeria plays host to well over 85% of these exploration and production activities. Hence, the severity of this threat caused by drilling wastes by itself cries out for urgent adequate remedial actions. Disposal of waste is usually in temporary pits awaiting final disposal or is in permanent disposal pits. Whichever the case, many a times overflows are witnessed into the surrounding arable land. These overflows carry along the full compliments of the toxic components of chemicals (some of which are toxic to even microorganisms) and oil [16].

Several research works have been carried out on the degradation of different organic compounds including all those present in the drill cuttings [17–19].

The ability of *Staphylococcus* sp. isolated from the soil to utilise an oil base and its potential for application in bioremediation process involving oil-based drilling muds have been demonstrated by Nweke and Okpokwasili [21]. The role of composition in the degradability and toxicity of drilling muds was examined, and the oil-based muds were observed to be relatively toxic to the ecosystems [2, 8]. Research has shown that drilling muds and its additives can be biodegraded by microorganisms [22] and their toxic effects can be either acute or chronic [2, 16, 19]. The effect of drilling chemicals on nitrate utilisation and logarithmic rate of growth of *Nitrobacter* was investigated by Okpokwasili and Odokuma [16], and the results showed that drilling chemicals inhibit an aspect of nitrification in the biosphere, thereby negatively affecting soil and water fertility. The biodegradation, mobility and photo toxicity of fuel oil hydrocarbons contained in drill cuttings were investigated in an agricultural soil and were found to modify the soil fertility [20]. Studies have also shown that oil spill dispersants and drilling fluids affect the ability of marine bacteria to metabolise substrates in the environment [23]. They were also observed to enhance the growth rate of *Bacillus* species and decrease that of *Staphylococcus* species with increasing concentrations of oil spill dispersants [24].

Researches have abundantly shown that naturally occurring microbial degradation mechanisms in the environment result in the biodestruction of toxic substances such as hydrocarbons [25, 26]. Microbial degradation often represents the most desirable form of attenuation because of the irreversible nature of the reaction. In the majority of cases, microbial degradation is a detoxifying mechanism, which leads to complete mineralisation [26]. Evidence from different investigations had suggested that microbial degradation of oil and petroleum in the environment is affected by limiting amounts of essential nutrients especially nitrate and phosphate salts [4, 26]. The limiting factors include restricted substrate assimilation capacity of the microbial population, presence of toxic substrate, optimal temperature and pH of the medium. These conditions greatly affect the rate at which the microbes function.

It had been observed that oil-based drill cuttings contain chromium (a pollutant heavy metal) at concentrations that could seriously threaten the life of edaphic systems, if it is continually injected into the ground without further remedial treatment measures [7]. The sorption of chromium by activated carbon from the flocculation effluent of liquid-phase oil-based drill cuttings (L-P-OBDC), which is heavily laden by oil/grease and suspended solids, had been investigated by Okparanma and Ayotamuno [9]. Toxicity and possible environmental impacts that may result from the indiscriminate disposal of drilling muds and cuttings derived from Ewan and Dibi offshore wells in the Niger Delta petroleum province of Nigeria had been studied by Adewole et al. [27]. Lethal and sublethal effects of the drilling-fluid XP-07 on the gill and liver morphology of *Tilapia guineensis* fry had been investigated in the laboratory using static bioassay for 96 h and 12 weeks exposure, respectively. At the end of 96 h, the gall and liver sections indicated minor lesions which were characterised by epithelial lifting and hyperplasia in gills, while irregularity in shape and size of the liver cell coupled with the presence of macrophages was observed in the liver [28].

## 2. Description of drill cuttings

Drill cuttings are the spoil which results from the exploration and production of well drilling activities and comprise small pieces of the strata through which the well is drilled. Together, they pass through the wellhead flow line into the primary shale shakers. The primary shale shakers are equipped with vibrating motor, and their vibrations make possible the filtration of drilling mud from drill cuttings through screen mesh on the shakers. The end-product of this filtration process is cutting with a much reduced drilling-fluid content. This system is placed downstream of the bell nipple, which is designed to minimise the contamination of cuttings due to mud, solids, caving and recirculated drill solids.

## 3. Types and compositions of drilling muds

The composition of drilling fluid depends upon the requirements of the particular drilling operation. The holes must be drilled through different types of formation requiring different types of drilling fluids. Economics, fluid contamination, available makeup water, pressure, temperature and many other factors are significant in the choice of drilling fluids [29]. A drilling fluid may be composed only of air. For example, one may decide to use air for drilling a hard rock of mountain size. Also water alone may be used to drill stable consolidated areas. However, in some areas, drilling can be started with water and the drilled solids incorporated into the water resulting in reasonably good mud. Generally, the functions to be performed require many properties, which cannot be obtained from ordinary liquid alone. Consequently, it may be necessary to add commercial clay to the water prior to drilling operations. Clay may also be added to oil to produce oil-based drilling mud. Both water and oil-based drilling muds are used in drilling oil wells.

Drilling muds are suspensions of solids (e.g. clay, barite, small cuttings) in liquid emulsions with chemical additives as required to modify their properties UKOOA [5]. Numerous chemical products are used daily in oilfield operations. The type and volume of chemicals vary depending on the type of production [6, 29]. Exploration and production operations generate chemicals such as crude oil, condensate, natural gas, hydrogen sulphide, carbon dioxide, heavy metals, brine salts and solid cuttings. These substances alongside with used muds and additives constitute what is called drilling wastes.

Drilling requires drilling mud to lubricate the drill bit, carry drill cuttings (rock chippings drilled from the reservoir formation) to the surface and control the downhole formation pressure of reservoir fluids. Water-based mud consists of natural clays and additives (organic or inorganic) to achieve proper density, viscosity and lubrication characteristic. Additives of particular concern from a pollution viewpoint are ferrochrome lignosulphonate (chromium pollution) and lead compounds (lead pollution). Oil-based muds contain oxidised asphalt, organic acid, alkali, stabilising agents and low toxic oil. Clay solids and weighing agents can also be added. Oil-emulsion muds are also used either as oil-in-water or water-in-oil varieties DPR [1]. Drilling muds contain heavy metals, either present in the mud for specific function (e.g. barium, chromium, zinc) or as impurities in natural mud additives (clay minerals and barite); such impurities include arsenic, copper, nickel and lead. Drilling mud also acquire heavy metal from the rock minerals in the drill cuttings and formation water [7]. These metals not only give rise to deterioration in environmental quality but also are potentially toxic to terrestrial plants [9]. The main sources of these elements are the two major solid components of drilling muds, which are barite and clays. Barium sulphate is the major constituent (80–90%) of barite, a weighting agent used to increase mud density. The amount of barite used depends on the drilling depth and the downhole condition and can be as high as 0.2 g barium per gramme of mud.

Several other potentially toxic metals such as cadmium and mercury are also present as impurities. Clay minerals like bentonite are added to water-based muds to enhance viscosity and suspension of properties. These clay minerals also act as fluid loss controllers. They constitute the second most abundant solid materials in water-based muds. Since clays are naturally occurring, their compositions largely depend on their geological source. Clay minerals also vary significantly in their potential for absorbing positively charged species DPR [1].

Drilling muds may also contain chromium in a variety of chemical forms. It is mostly complexed with lignosulphonate material as the hexavalent cation and accounts typically for 3–4% by weight of lignosulphonate complex. Lignosulphonates control viscosity in water-based muds by acting as thinning agents or deflocculants for clay particles. Calcium and iron compounds combine with chromium to form common ferrochrome lignosulphonates.

One of the major components of oil-based mud is the base oil. Mineral oil used in drilling muds is mixtures of alkanes, alkenes, cycloalkanes and aromatic hydrocarbons produced by petroleum refining DPR [1]. Diesel oil and lower toxicity base oil are produced initially from the kerosene and light gas oil fractions of crude oil. The aromatic content of diesel oil is usually in the range of 20–30%. Lower toxicity base oils are produced by further treatment by base stocks and are usually clear or slightly yellow with little odour. They normally consist of a narrow distillation cut of midrange aliphatic materials with an aromatic carbon content of 1–10%, although some are blended for certain application.

Diesel oil was used in the initial oil-based mud formulations, but they were found to be toxic on a wide range of microorganisms. Due to the higher concentration of the potentially toxic materials (2-, 3- and 4-ring aromatics) in diesel, a large number of highly refined mineral oils with a much smaller percentage of aromatic compounds than diesel subsequently became an alternative. This lower toxicity or alternative base oil is a mixture of alkane or cycloalkanes, and they are referred to as the synthetic-based muds.

Synthetic-based muds are classified according to molecular structure of the synthetic-based fluids which can be esters, ethers, etc. They have drilling and operational properties similar to oil-based muds but have the advantage of being more environmentally friendly. However, environmental monitoring of sites where cuttings from synthetic-based mud drilled well was discharged to the sea indicated that the majority of synthetic mud system fluids were in fact not as biodegradable in the marine environment as the laboratory tests had indicated UKOOA [5].

Regardless of the type of complexity of the drilling fluid or mud, nine basic functions stand out glaringly for consideration. These are to remove drilled cuttings from the hole; control subsurface pressure; cool and lubricate the bit and drilling stem; minimise washout and damage to well bore; suspend cuttings, weight materials and others solids when circulation is stopped; and help support the weight of drill stem and casing.

## 4. Guidelines and standards of the regulatory authority

Drilling wastes, as described in the guidelines and standards of the regulatory authority in Nigeria, the Department of Petroleum Resources, are forbidden pretreatment discharge into the environment without first ascertaining the nil or minimum impacts via carrying out the environmental impact assessment and environment evaluation report studies.

Even though the vast majority of exploration and production wastes and contaminated soil are not hazardous, the public and regulators alike have expressed concern regarding the environmental impact of exploration and production wastes on the environment DPR [1].

Specific chemicals considered hazardous to the environment and whose presence and level must be ascertained prior to disposal of drilling wastes include sulphur (iv) oxide, hydrogen tetraoxosulphate (vi) acid, hydrogen sulphide, ammonia, chloride, formaldehyde, hydrogen fluoride, crude oil (both on its own and as drill cuttings) natural gas, fuel oil, condensate, calcium hydroxide, hydrochloric acid, amines, propane, ethylene, glycol, methanol and heavy metals (zinc, chromium, cadmium, etc.). In addition to these possible hazardous drilling waste constituents, some whole field chemicals also contain hazardous constituents. These whole chemicals include oil-based mud additives, solvents, fuels, herbicides, biocides, emulsion breakers, corrosion inhibitors, deflocculants, defoamers, scale inhibitors, pH buffers, cement additives and weighting materials [29].

## 5. Modes of toxicity

As a consequence of the high demand for rapid, inexpensive and relatively simple screening tests for evaluating the acute toxicity of chemicals in the environment, the need for the knowledge of the mechanisms of toxicity received increasing attention. Chemical agents, toxic to microbial population, act by either inhibition or destruction of cellular components vital to cell functions.

Exposures of organisms to heavy metals result in the occurrence of a variety of abnormalities such as interference with cell wall synthesis, decreased enzyme activity and inactivation of DNA and RNA [30]. Other agents responsible for membrane disruption include quaternary ammonium compounds. These compounds also inhibit bacterial oxidase and dehydrogenase systems causing protein denaturation and enzyme suppression [31].

## 6. Test organisms

The prime considerations in selecting test organisms for toxicity bioassay are their sensitivity to the factors under consideration; their geographical distribution, abundance and availability within a practical size range throughout the year; their recreational, economic and ecological importance; the availability of culture methods for rearing them in the laboratory and a knowledge of their requirements; and their general physical conditions and freedom from parasites and diseases APHA [32]. Macro-organisms like fish, rats, snails, crabs and prawns are used in toxicity test because of certain feature they possess. These include easy to rear, visible to sight, count, notice and detect, cheap to procure.

However, the advantages of bacteria over macro-organisms include the fact that bacteria take less time to grow, multiply into millions within a short period so that the process of getting bacteria for biomonitoring is simple, sensitive, rapid and inexpensive. The use of microorganisms in bioassays is based on cell lysis and the inhibition of physiological processes such as respiration [33].

## 7. Biodegradation

Biodegradation has been defined as the gradual breakdown of a compound to its constituents by microorganisms. In other words, it is the natural process whereby microorganisms use up such substances as energy source, which are broken down into other similar substances such as fatty acids, carbon dioxide and water, thereby multiplying in number in the process [34, 35]. The microbial utilisation of wastes is based primarily on the natural degradative capabilities of microorganisms. These capabilities are derived from the metabolic diversity of both bacteria and fungi, which evolved from their role in the biochemical cycling of organic and inorganic compounds in the environment. The presence of hydrocarbons in the environment selects hydrocarbon-utilising microorganisms within the total heterotrophic population [26, 36], irrespective of whether they are contaminated or not. Indeed, all available evidences suggest that hydrocarbon utilisation at contaminated sites proceeds naturally.

Biodegradation often represents the most desirable form of detoxification because of the irreversible nature of the reaction [15]. For the natural cleansing of hydrocarbon in the environment (soil/water), microorganisms are considered to play the major role. Oil-utilising microorganisms have been reported to include Gram-positive and Gram-negative organisms, and notable genera include *Pseudomonas*, *Alcaligenes*, *Rhodococcus*, *Acinetobacter*, *Arthrobacter* and *Corynebacterium* [35]. Odjadjare et al. [34] had earlier reported that for biodegradation process to be considered an acceptable remediation option, there must be evidence that the microorganisms were responsible for the degradation of organic contaminant. These evidences may include stepwise growth patterns of elevation, biomass concentration and metabolic changes over time.

Abiotic factors have been observed to influence both the weathering and degradation of petroleum spilled in the environment. Atlas [37] observed that after the removal of low-to-medium weight molecules by artificial weathering, weathered light oils were biodegraded by bacteria at a higher rate and to a greater extent than weathered heavy oils and fresh oils. Crude oils therefore contain some factors that are toxic to microorganisms. The complex high-molecular-weight compounds of heavy oils may also be resistant to degradation. Okpokwasili and Okorie [38] assessed the biodegradability of used and unused lubricating oil using mixed culture of bacteria isolated from used oil. Total viable counts and analysis by chromatography confirm that used oil has been modified in service and is more degradable than unused oil.

The susceptibility of petroleum hydrocarbons to utilisation is also determined by the structure, configuration and molecular weight of the hydrocarbon molecule [36, 37]. Despite the abundance of hydrocarbon-degrading microorganisms, the successful degradation of hydrocarbons requires optimization of environmental conditions (nutrients, temperature and pH). Suitable growth temperature and available supplies of fixed forms of nitrogen, phosphorus and molecular oxygen are required so that bioremediation occurs at the maximum rate and to the greatest possible extent. Atlas [37] reported that inoculation of polluted areas with oil-decomposing microorganisms was ineffective because of growth-limiting nitrogen and phosphorus concentrations in sea water.

There have been several reports showing that biodegradation of hydrocarbons that have low solubility or that may be absorbed by soil particles can be enhanced by the addition of biosurfactants [39]. Biosurfactants increase the availability of hydrocarbons by increasing hydrocarbon solubility and desorption. It also influences surface properties of degrading cells, resulting in enhanced hydrocarbon utilisation. Bioremediation has actually become an accepted technology for restoration of contaminated environments. However, successful applications have primarily involved degradable organics. Bioremediation is used infrequently with more recalcitrant pollutants, often because microorganisms indigenous to contaminated environments lack appropriate degradative capabilities [35]. In these cases, it may be possible to enhance bioremediation by adding microorganisms that have catabolic functions to degrade polluted site [40]. This process is referred to as bio-augmentation. Augmentation of the natural microbial population is by genetic engineering process and is an active area of research for recalcitrant contaminants and other commercial products.

It had been observed that pure cultures of individual species have only limited substrate range and are of little help in consuming the complex hydrocarbon mixtures found in crude oil [41]. Most effort in oil spill clean-up by microbial seeding has been directed towards the use of mixed culture, such that the individual species of the consortium would have the ability to consume either aliphatic, aromatic or polynuclear aromatic hydrocarbons [42, 43]. During the growth of mixed culture, interactions such as the production by one strain of organisms, toxic compounds that are inhibitory to others and the differences in growth rates among species can limit the number of surviving organisms. An ideal solution for rapid microbial consumption of crude oil for the rapid removal of oil slick will be to construct novel strains, which would have the genetic potentialities to degrade simultaneously, a variety of oil components.

Recombinant DNA technology, however, permits the incorporation of the diverse types of genetic information extracted from several organisms into a single organism [37]. Through genetic engineering a "super bug" has been created that is capable of degrading many different hydrocarbons structures and that is potentially useful in oil pollution abatement programme [37].

The biodegradation of oil pollutants is not a new concept. It has, however, taken a new significance as an increasingly effective and potentially inexpensive clean-up technology. Its potential contribution as countermeasure biotechnology for decontamination of oil-polluted ecosystems is enormous. The successful treatment of spilled oil requires both a proven microbial population and well-engineered support systems that optimise the microbes' inherent favourable characteristics while providing needed protection against the catalysts.

## 8. Drill cutting management

The exploration and production industries continue to be faced with the challenges associated with discovering and economically recovering new oil and gas reserves. Explorations of new fields, as well as the exploitation of existing fields, require well drilling operations, along with the continuous task of reducing drilling programme cost intelligent and environmentally prudent disposal solutions for the associated drilling fluids and cuttings. Drilling waste management is, therefore, the planning and implementation of a cautious waste collection, treatment and disposal plan. Within the planning, detailed analysis of the treatment and disposal options are performed.

The disposal options listed below are typically used:

**Injection:** In the injection process, the cuttings are converted into slurry, which is then pumped into a receiving formation at pressures exceeding the fracture gradient.

**Offshore discharge:** Depending on local regulations, cuttings with low environmental risk can be disposed into the sea. However, most local regulations require relative levels pretreatment of waste to reduce the concentration of oil-on-cuttings to <6.9% on a wet weight basis.

**Solidification:** This is considered as a nontreatment option. The waste is mixed with a material (activated lime, kiln dust, fly ash, cement, etc.) to form a solid product that immobilised potential contaminants.

**Waste-soil mixtures:** Waste-soil mixtures are considered a nontreatment option. It consists of mixing the waste with soil or subsoil to decrease the concentration of the potential contaminants. The chemical proportion of the waste is adjusted to meet regulatory standards.

**Thermal desorption:** This process is a separation and recovery process that is achieve by the application of heat which results in three streams of water, oil and solids. The vapours are cooled and separated into water and oil phases. The recirculating processes usually include a water phase which is used to cool and return moisture to the solid stream, the oil phase is recovered and used as a drilling-fluid system or fuel source, and the solids could be disposed of or reused.

**Bioreactor:** This process involves the application of indigenous topsoil bacteria and fungi with the ability to utilise the base oil as a primary source of carbon. The rate of biodegradation of the cuttings will be faster if properly mixed with water, oxygen and appropriate nutrients. In a bioreactor system, factors that should be properly monitored are aeration, mixing, nutrients and bacterial colony strength.

**Land farming:** In this technique, the cuttings are applied to a soil surface and ploughed to ensure adequate mixing and aeration. The rate of degradation is controlled by using the correct levels of nutrients, soil moisture and air.

## 9. Physicochemical qualities of drill cuttings collected from Ologbo oilfield wells

The physicochemical characteristics of drill cuttings collected from Ologbo oilfield wells are shown in **Table 1**. The pH ranges from 5.2 to 5.9, the electrical conductivity ranges from 220.0 to 309.0 μS, the oil and grease range from 38.0 to 309.0 mg/kg, and the total hydrocarbon content was 33.5–79.1 mg/kg. The variations in the parameter values obtained in this study clearly attest to the different characteristics of the drill cutting profiles from oil and gas wells. This evaluation had further buttressed the fact that drill cuttings are usually laden with high concentrations of oil and grease, especially when oil-based or synthetic-based muds are used in the drilling process. At these high concentrations, discharge of drill cuttings into the environment without pretreatment can have adverse effects on the edaphic systems and on groundwater if it permeates through the aquifer and also as runoff into water bodies [2].

| | | Cutting depth (m) | | | Acceptable limits |
|---|---|---|---|---|---|
| | | 0–305 | 610–915 | Composite | DPR |
| Location 1 | pH | 5.2 | 5.2 | 5.2 | 6.5–9.0 |
| | Electrical conductivity (μS/cm) | 231.0 | 309.0 | 274.0 | 8 mm hos/cm |
| | Oil and grease (mg/kg) | 38.0 | 125.0 | 175.5 | 100 mg/kg |
| | Total hydrocarbon content (mg/kg) | 37.6 | 57.1 | 59.0 | 10 mg/kg |
| Location 2 | pH | 5.8 | 5.3 | 5.3 | 6.5–9.0 |
| | Electrical conductivity (μS/cm) | 220.0 | 278.0 | 253.0 | 8 mm hos/cm |
| | Oil and grease (mg/kg) | 40.0 | 252.0 | 258.0 | 100 mg/kg |
| | Total hydrocarbon content (mg/kg) | 33.5 | 43.4 | 47.2 | 10 mg/kg |
| Location 3 | pH | 5.3 | 5.9 | 5.8 | 6.5–9.0 |
| | Electrical conductivity (μS/cm) | 222.0 | 262.0 | 279.0 | 8 mm hos/cm |
| | Oil and grease (mg/kg) | 48.0 | 289.1 | 309.0 | 100 mg/kg |
| | Total hydrocarbon content (mg/kg) | 52.5 | 78.0 | 79.1 | 10 mg/kg |

*Over all mean value source: DPR [1], Imarhiagbe [44].*

**Table 1.**
*Some physicochemical parameters of drill cuttings from Ologbo oilfield wells.*

## 10. Heavy-metal qualities of drill cuttings collected from Ologbo oilfield wells

Several studies have shown that waste drill cuttings harbour levels of heavy metals (Veritas (2000), [6, 45]). Study by Imarhiagbe [44] revealed that drill cutting

| | | Cutting depth (m) | | | Standards |
|---|---|---|---|---|---|
| | | 0–305 | 610–915 | Composite | DPR |
| Location 1 | Pb | 2.6 | 1.2 | 1.5 | 0.0 |
| | Fe | 4.4 | 2.7 | 2.2 | 0.3 |
| | Zn | 0.8 | 2.0 | 1.5 | 5.0 |
| | Ni | 1.9 | 1.5 | 1.7 | 0.0 |
| | Cu | 0.0 | 0.0 | 0.0 | 1.3 |
| | Cr | 0.0 | 0.0 | 0.0 | 0.1 |
| Location 2 | Pb | 0.8 | 0.9 | 0.7 | 0.0 |
| | Fe | 2.4 | 2.0 | 2.0 | 0.3 |
| | Zn | 2.1 | 1.7 | 2.0 | 5.0 |
| | Ni | 1.2 | 1.0 | 1.2 | 0.0 |
| | Cu | 0.0 | 0.0 | 0.0 | 1.3 |
| | Cr | 0.0 | 0.0 | 0.0 | 0.1 |

| | | Cutting depth (m) | | | Standards |
|---|---|---|---|---|---|
| | | 0–305 | 610–915 | Composite | DPR |
| Location 3 | Pb | 0.5 | 0.5 | 0.5 | 0.0 |
| | Fe | 3.8 | 3.3 | 2.9 | 0.3 |
| | Zn | 0.0 | 0.2 | 0.0 | 5.0 |
| | Ni | 0.8 | 1.2 | 1.2 | 0.0 |
| | Cu | 0.0 | 0.0 | 0.0 | 1.3 |
| | Cr | 0.0 | 0.0 | 0.0 | 0.1 |

*Over all mean value source: DPR [1], Imarhiagbe, [44].*

**Table 2.**
*Levels of heavy metals of drill cuttings from Ologbo oilfield wells.*

samples from Ologbo oilfield wells contained high concentrations of iron compared to other heavy metals such as lead, zinc, nickel, copper and chromium (**Table 2**). However, the levels of heavy metals were less than the permissible limits as set by the Department of Petroleum Resources, which is the Nigeria government regulator of the oil and gas sector [1]. The presence of these toxic heavy metals may play an inhibitory role in the pollutant's biodegradation through the interaction with microbial enzymes [46]. Therefore, preventing heavy-metal pollution is very critical, considering the difficulty and cost of cleaning contaminated environment [47].

## 11. Microbiological qualities of drill cuttings collected from Ologbo oilfield wells

The order of microbial population density of composite cutting samples from oilfield wells ranges from 10 to $10^5$ cfu/g for total heterotrophic bacterial counts,

| | | | Cutting depth (m) | | |
|---|---|---|---|---|---|
| | | | 0–305 | 610–915 | Composite |
| Location 1 | Total heterotrophic bacterial counts | | $7.2 \times 10^3$ | $1.1 \times 10$ | $7.23 \times 10^5$ |
| | Total heterotrophic fungal counts | | $3.0 \times 10^3$ | 0.0 | $3.3 \times 10^3$ |
| | Total heterotrophic anaerobic bacterial counts | | $1.7 \times 10^2$ | $5.3 \times 10^3$ | $5.7 \times 10^3$ |
| | Total mud utilizing Bacterial count | WBM | $5.0 \times 10$ | $2.5 \times 10$ | $7.7 \times 10$ |
| | | NABM | $2.0 \times 10$ | 0.0 | $3.2 \times 10$ |
| | Total mud utilizing fungal count | WBM | $2.8 \times 10$ | $1.0 \times 10$ | 0.0 |
| | | NABM | $1.3 \times 10$ | 0.0 | 0.0 |
| Location 2 | Total heterotrophic bacterial counts | | $5.4 \times 10^3$ | 0.0 | $5.4 \times 10^5$ |
| | Total heterotrophic fungal counts | | $2.7 \times 10^3$ | 0.0 | $3.7 \times 10^5$ |
| | Total heterotrophic anaerobic bacterial counts | | $1.5 \times 10^2$ | $2.7 \times 10^3$ | $4.4 \times 10^3$ |
| | Total mud utilizing bacterial count | WBM | $4.2 \times 10$ | $2.0 \times 10$ | $6.5 \times 10$ |
| | | NABM | $1.8 \times 10$ | $1.0 \times 10$ | $4.0 \times 10$ |
| | Total mud utilizing fungal count | WBM | $2.3 \times 10$ | $1.1 \times 10$ | 0.0 |
| | | NABM | $1.1 \times 10$ | 0.0 | 0.0 |

| | | | Cutting depth (m) | | |
|---|---|---|---|---|---|
| | | | **0–305** | **610–915** | **Composite** |
| Location 3 | Total heterotrophic bacterial counts | | $2.8 \times 10^3$ | 0.0 | $4.0 \times 10^5$ |
| | Total heterotrophic fungal counts | | $2.1 \times 10^3$ | 0.0 | $2.5 \times 10^5$ |
| | Total heterotrophic anaerobic bacterial counts | | $1.5 \times 10^2$ | $4.1 \times 10^3$ | $3.8 \times 10^3$ |
| | Total mud utilizing bacterial count | WBM | $4.5 \times 10$ | 0.0 | $6.0 \times 10$ |
| | | NABM | $2.2 \times 10$ | 0.0 | $2.5 \times 10$ |
| | Total mud utilizing fungal count | WBM | $1.3 \times 10$ | 0.0 | 0.0 |
| | | NABM | $1.0 \times 10$ | 0.0 | 0.0 |

*Over all mean values; WBM, water-based mud; NABM, nonaqueous-based mud. Source: [44].*

**Table 3.**
*Mean total heterotrophic microorganism counts (cfu/g) of drill cuttings from Ologbo oilfield wells.*

| Bacterial isolates | Fungal isolates |
|---|---|
| *Enterobacter* spp. | *Aspergillus niger* |
| *Micrococcus* spp. | *Aspergillus clavatus* |
| *Bacillus* spp. | *Rhizopus nigricans* |
| *Staphylococcus* spp. | *Penicillium expansum* |
| *Clostridium* spp. | *Penicillium glaucus* |
| *Mycobacterium* spp. | *Cladosporium* spp. |
| *Desulfotomaculum* spp. | |
| *Nocardia* spp. | |
| *Citrobacter freundii* | |

*Source: [44].*

**Table 4.**
*Microorganisms isolated from drill cuttings collected from Ologbo oilfield wells.*

total heterotrophic fungal counts, total heterotrophic anaerobic bacterial counts and total utilising bacterial counts, respectively (**Table 3**).

Microorganisms associated with drill cuttings from Ologbo oilfield wells are stated in **Table 4**. Gram-positive bacteria such as *Clostridium* spp., *Mycobacterium* spp., *Bacillus* spp., *Staphylococcus* spp., *Nocardia* spp. and *Micrococcus* spp. were the predominant bacterial isolates. The Gram-negative bacterial isolates were *Desulfotomaculum* spp., *Citrobacter freundii* and *Enterobacter* spp. The predominant fungal isolates were *Cladosporium* spp., *Penicillium glaucus*, *Penicillium expansum*, *Rhizopus nigricans*, *Aspergillus clavatus* and *Aspergillus niger*.

## 12. Conclusion

In conclusion, it was quite clear that drill cuttings emanating from these oilfield locations were unsafe for disposal, and the regulating agency in Nigeria needs to enforce treatment prior to disposal; otherwise appropriate sanctions should be applied.

Recommendations from the study of drill cuttings from Ologbo oilfield wells:

I. It is therefore recommended that oil exploration industries should adhere strictly to the procedures and instructions described for waste management in oil drilling exploration and procedure waste management in Nigeria.

II. The ban on the use of toxic oil-based mud by oil exploration companies in Nigeria and globally should be properly monitored and enforced by appropriate regulatory agencies.

III. It is therefore recommended that wastes resulting from oil activities should be properly treated before it is disposed into the receiving environment.

IV. The new trend is that waste materials from the oil industries should be seen as raw materials for reuse. If solid wastes are properly treated, it can serve as raw materials for cement-producing plants, bricks and expanded clay-producing plants and can also be used in land restoration projects. Appropriate application of these technologies can help in the creation of jobs for the teeming youths where these projects are sited.

V. In furtherance to enforcement, disposal of cuttings and drilling wastes into the onshore environment should be totally discouraged and litigated. This act had observed to have significant adverse effect on the aquatic ecosystem.

## Author details

Emmanuel Esosa Imarhiagbe* and Nosa Omoregbe Obayagbona
Department of Environmental Management and Toxicology, Faculty of Life Sciences, University of Benin, Benin City, Nigeria

*Address all correspondence to: esosa.imarhiagbe@uniben.edu

## References

[1] Department of Petroleum Resources (DPR). Environmental Guidelines and Standards for the Petroleum Industry in Nigeria (EFAPSIN). 2002. p. 313

[2] Odokuma LO, Ikpe MD. Role of composition on the degradability and toxicity of drilling muds. African Journal of Applied Zoology and Environmental Biology. 2003;**5**:6-13

[3] Soegianto A, Irawan B, Affandi M. Toxicity of drilling waste and its impact on gill structure of post larvae of Tiger prawn (*Penaeus monodon*). Global Journal of Environmental Research. 2008;**2**(1):36-41

[4] Imarhiagbe EE, Atuanya E. Aerobic degradation of drill muds by axenic and mixed bacterial isolates from drill cuttings at Ologbo, Edo state, Nigeria. Nigerian Journal of Biotechnology. 2013;**25**:46-52

[5] United Kingdom Offshore Operators Association (UKOOA). Methodology for the evaluation of management and disposal option for drill cuttings on the seabed. In: UKOOA Drill Cutting Initiative, Research and Development Dames and Morre; 2000. p. 85

[6] Gbadebo AM, Taiwo AM, Eghele U. Environmental impacts of drilling mud and cutting wastes from the Igbokoda onshore oil wells, southern Nigeria. Indian Journal of Science Education and Technology. 2010;**3**(5):504-510

[7] Ayotamuno JM, Okparanma RN, Araka PP. Bioaugmentation and composting of oil-field drill-cuttings containing polycyclic aromatic hydrocarbons (PAHs). Journal of Food, Agriculture and Environment. 2009;**7**(2):658-664

[8] Ekpo M, Ekanem AIJ. Toxicity of petroleum drilling mud additive (chrome lignosulfonate) on micro-organisms and fresh water fish tilapia (*Oreochromis nilotieus*). Nigerian Journal of Microbiology. 2000;**14**(2):55-60

[9] Okparanma RN, Ayotamuno M. Predicting chromium (vi) adsorption rate in the treatment of liquid-phase oil-based drill cuttings. African Journal of Environmental Science and Technology. 2008;**2**(4):68-74

[10] Frost BR. A cretaceous niger delta petroleum system in extended Abstracts. AAPG/ABGP Hedberg Research Symposium. Petroleum Systems of the South Atlantic Margin. Brazil: Rio de Janeiro; November 16-19 1997

[11] Paul IA. A historical perspective of petroleum on Nigeria's economic crisis since independence. Global Journal of Human-Social Science. 2015;**15**(2):17-24

[12] Ameloko AA, Omali AO. Reservoir characterization and structural interpretation of seismic profile: A case study of z-field, Niger Delta, Nigeria. Petroleum and Coal. 2013;**55**(1):37-43

[13] Ekweozor CM, Daukoro EM. Northern delta depobelt portion of the Akata-agbada petroleum system, Niger-delta, Nigeria. In: Magoon LB, Dow WG, editors. The Petroleum System—From Source to Trap. Vol. 60. Tulsa: American Association of Petroleum Geologists; 1994. pp. 599-614

[14] Odokuma LO, Okpokwasili GC. Role of composition in the bio-degradation of oil spill dispersants. Waste Management. 1992;**12**:39-43

[15] Nwaugo, V. O; Onyeagba, R.A., Chima, G. N and. Agwaranze, D. I. (2007). Effects of drilling wastes on the farmland microbial spectrum in Egbema, south-south Nigerian community. Research Journal of Microbiology 2: 632-638

[16] Okpokwasili GC, Odokuma LO. Response of nitrobacter to toxicity of hydrocarbon fuels. Journal of Petroleum Science and Engineering. 1996;**16**:89-93

[17] Okoro C. Aerobic degradation of synthetic based drilling mud base fluids by gulf of Guinea sediments under natural environmental conditions. Life Science Journal. 2011;**8**(2):569-576

[18] Shaeyan M, Tirandaz H, GhanbarpourS,SeyedipourN,ShavandiM, Dastgheib SMM. Bioremediation of a drilling waste-contaminated soil; biotreatability assessment and microcosm optimization for developing a field-scale remediation process. Iranian Journal of Biotechnology. 2018;**16**(3):193-199

[19] Wu Y, Bai J, Zhong K, Huang Y, Qi H, Jiang Y, et al. Antibacterial activity and membrane-disruptive mechanism of 3-p-trans-coumaroyl-2-hydroxyquinic acid, a novel phenolic compound from pine needles of *Cedrus deodara*, against *Staphylococcus aureus*. Molecules. 2016;**21**(8):1-12

[20] Chaineau CH, Morel JL, Oudot J. Land treatment of oil-based drill cuttings in an agricultural soil. Journal of Environmental Quality. 1996;**25**:858-867

[21] Nweke CO, Okpokwasili GC. Drilling fluid base oil biodegradation potential of a soil *Staphylococcus* species. African Journal of Biotechnology. 2003;**2**(9):293-295

[22] Enemchukwu E, Okpokwasili GC. Biodegradability of drilling mud additives. Nigerian Journal of Microbiology. 2003;**17**(1):1-6

[23] Okpokwasili GC, Nnubia C. Effects of oil spill dispersants and drilling fluids on substrate specificity of marine bacteria. Waste Management. 1996;**15**(7):515-520

[24] Nnubia C, Okpokwasili GC. The microbiology of drilling mud cuttings from a new off-shore oil field in Nigeria. Environmental Pollution. 1993;**82**:153-156

[25] Atagana HI. Bioremediation of creosote contaminated soil: A pilot-scale land farming evaluation. World Journal of Microbiology and Biotechnology. 2003;**19**:571-581

[26] Marinescu M, Dumitru M, Lăcătuşu A. Biodegradation of petroleum hydrocarbons in an artificial polluted soil. Research Journal of Agricultural Science. 2009;**41**(2):157-162

[27] Adewole GM, Adewale TM, Ufuoma E. Environmental aspect of oil and water based drilling muds and cuttings from Dibi and Ewan off-shore wells in the Niger-Delta, Nigeria. African Journal of Environmental Science and Technology. 2010;**4**(5):284-292

[28] Vincent-Akpu IF, Sikoki FD, Utibe D. Toxicity of drilling fluid XP-07 to *Tilapia guineensis* fry. African Health Sciences. 2010;**9**(2):68-76

[29] Martin WB. Oil chemicals and regulatory concerns. Journal of Petroleum Science and Engineering. 1991;**35**:14-16

[30] Izah SC, Inyang IR, Angaye TCN, Okowa IP. A review of heavy metal concentration and potential health implications of beverages consumed in Nigeria. Toxics. 2017;**5**:1-15

[31] Nester EW, Roberts CE, Pearsall NN, Anderson DG, Nester MT. Environmental microbiology. In: Microbiology: A Human Perspective. USA: The McGraw-Hill Companies, Inc.; 1998. 848 pp

[32] American Public Health Association (APHA). Standard Methods for the

Examination of Water and Waste Water. 20th ed. Washington, DC: American Public Health Association; 1998. 1571 pp

[33] Yao L, Naeth MA, Jobson A. Soil microbial response to waste potassium silicate drilling fluid. Journal of Environmental Sciences. 2015;**29**:189-198

[34] Odjadjare EEO, Ajisebutu SO, Igbinosa EO, Aiyegoro OA, Trejo-Hernandez MR, Okoh AI. Escravos light crude oil degrading potentials of axenic and mixed bacterial cultures. The Journal of General and Applied Microbiology. 2008;**54**:277-284

[35] Prescott LM, Harley JP, Klen DA. Microbiology. 6th ed. UK: McGraw Hill Companies Inc.; 2005. 962 pp

[36] Nilanjana D, Preethy C. Microbial degradation of petroleum hydrocarbon contaminants: An overview. Biotechnology Research International. 2011;**2011**:1-13

[37] Atlas RM. Microbiology, Fundamentals and Applications. 2nd ed. New York: Macimillan Publishing Company; 1988. 563 pp

[38] Okpokwasili GC, Okorie BB. Biodeterioration potentials of micro-organisms isolated from car engine lubricating oil. Tribology International. 1988;**21**(4):215-220

[39] Churchill SA, Griffin RA, Jones P, Churchill PF. Biodegradation rate enhancement of hydrocarbons by an oleophilic fertilizer and a rhamolipid biosurfactant. Journal of Environmental Quality. 1995;**24**:19-28

[40] Halden RU, Tepp SM, Halden BG, Owyer DF. Degradation of 3-phenoxybenzoic acid in soil by *Pseudomonas alcaligenes* POB 310 (pPOB) and two modified *Pseudomonas* strains. Applied and Environmental Microbiology. 1999;**65**:3354-3359

[41] Okoh AI. Biodegradation of bonny light crude oil in soil microcosm by some bacterial strains isolated from crude oil flow stations saver pits in Nigeria. African Journal of Biotechnology. 2003;**2**(5):104-108

[42] Ilori MO, Obayori OS, Adebusoye SA, Abe FO, Oyetibo GO. Degradation of aroclor 1221 by microbial populations of the Lagos lagoon. African Journal of Biotechnology. 2007;**6**(20):2369-2374

[43] Okerentugba PO, Ezeronye OU. Petroleum degrading potentials of single and mixed microbial cultures isolated from rivers and refinery effluent in Nigeria. African Journal of Biotechnology. 2003;**2**(9):288-292

[44] Imarhiagbe EE. Microbiology and biodegradability of drill mud cuttings at Ologbo, Edo State [PhD dissertation]. Benin City-Nigeria: Department of Microbiology; University of Benin; 2012. p. 186

[45] Joel OF, Amajuoyi CA. Determination of selected physicochemical parameters and heavy metals in a drilling cutting dump site at Ezeogwu-Owaza, Nigeria. Journal of Applied Sciences and Environmental Management. 2009;**13**(2):27-31

[46] Olaniran AO, Balgobind A, Pillay B. Bioavailability of heavy metals in soil: Impact on microbial biodegradation of organic compounds and possible improvement strategies. International Journal of Molecular Sciences. 2013;**14**(5):10197-10228

[47] United States Department of Agriculture (USDA). Natural Resources Conservation Service (NRCS): Heavy Metal Soil Contamination. Soil Quality Urban Technical Note No. 3. 411 S. Donahue Dr. Auburn, AL: Soil Quality Institute; 2000

Chapter 4

# An Improved Semi-Analytical Approach for Predicting Horizontal and Multilateral Well Performance

*Adesina Fadairo, Gbadegesin Adeyemi, Temitope Ogunkunle, Ayotomiwa Evbogame and Adedapo Adesina*

## Abstract

Field development and economic evaluation of hydrocarbon demand for an accurate model for predicting horizontal well performance as horizontal and multilateral wells have become far more prominent in the industry than vertical wells. Several approaches for modelling horizontal well performance have been studied and reported in the literature. Analytical approach is the easiest with large inaccuracy in the prediction of the horizontal well performance because of inability to apply it in reservoir-wellbore coupling equation. Numerical approach is more reliable for field application than analytical approach. However, it involves iterative nature that requires longer computational times. Semi-analytical approach is simpler and sufficiently exact for field applications if the governing fundamental flow equation is accurately modelled. This study presents a new semi-analytical model for predicting horizontal and multilateral well performance, which includes friction, acceleration and accumulation induced pressure drop along horizontal well length into the governing fundamental flow equations. The outcomes of the proposed model have been validated by field data gotten from gauge rate of 5660stb/d at steady-state condition. The estimated steady flow rate of 5593.9 stb/day obtained from the new approach shows an error of 1.2% which is seen to be more accurate than steady flow rate values obtained by four previous models that exhibited higher percentage errors when compared to gauge reading.

**Keywords:** pressure due to accumulation, pressure due to friction, horizontal well, multilateral well, well performance

## 1. Introduction

As a sequel to advancement in drilling and completion technology, there has been increasing interest in horizontal wells. Production enhancement and economic increment of hydrocarbon recovery have given horizontal wells completion advantages over vertical wells most especially in small and marginal reservoirs [1–5]. However, horizontal well is costlier to drill and complete than vertical well. With current innovation in technology, the petroleum industry has generally moved to horizontal wells, as it is fast becoming the traditional practice [1–10]. Multilateral wells display the same benefits that horizontal wells also do, as well as they can recover hydrocarbon simultaneously from more than one reservoir; this offers significant increments in well planning and economics [7–11].

There have been a few endeavours to predict horizontal well performance; these have led to the development of various models that describe the performance of horizontal and multilateral wells. Previously developed work has been done for the estimation of productivity, and they have all made assumptions that either the well allows for infinite conductivity or the flow along the length of the well is uniform. This assumption leads to the pressure drop along the well to be neglected, and hence it is assumed to be constant throughout the well length. However, it is not a practical assumption as it does not capture the reality of horizontal wells, particularly in long horizontal drain hole where the pressure drop along the length of the well is large and cannot be treated as the reservoir-to-wellbore pressure drop system of the vertical well [3].

Some authors have attempted a coupling model that accounts for wellbore flow, as well as reservoir inflow to estimate the performance of a single phase horizontal well at the point when the pressure drop in the wellbore becomes significant. Dikken [4] was one of the first experts to couple fluid flow in the lateral of the wellbore to the reservoir in-flow using a model; afterward, several models have been reported. The study demonstrated that in most practical circumstances, a wellbore exhibits flow either in the turbulent flow regime or transition flow regime into the wellbore and no laminar flow is present. Landman [5] further proposed enhancements to the model developed by Dikken by varying the productivity index (PI) along the wellbore, and the variations are due to changes in perforation density, permeability and the characteristics of the flow along the well. In the model, a method for evaluating the optimum perforation density results in specific inflow along the well length. Novy [12] generalized the work done by Dikken by developing a model which could be applied to single phase oil flow and gas flow. To handle the gas system, non-Darcy flow term was introduced to the equation by the author. Ozkan and Haciislamoglu [13] examined the impact of pressure drop inside the horizontal section and how a horizontal well responds to it. As such, they presented a general, semi-analytical model which couples reservoir inflow and wellbore flow hydraulics. They defined groups to correlate the response of horizontal well and how these are affected by wellbore hydraulics. Basically, pressure distributions and flux distribution along the lateral of the well were investigated, and they discussed the validity of the assumptions of infinite conductivity. Penmatcha et al. [14] investigated the need to optimise the well length and how it affects the drop in pressure along the horizontal well. They proposed that as the length of the horizontal well increases, there is more accessibility to larger contact with the reservoir; however, this also leads to an increase in resistance to flow, which many times reduces productivity. Ouyang et al. [9] developed a single-phase wellbore-flow model in their research that combined pressure drop due to acceleration, gravity and friction. They developed a model that was very applicable with distinct configurations of perforation at the wellbore and completions; the model developed could be used analytically with any model that describes inflow of fluids into the reservoir or used with reservoir simulations [15].

Chen et al. [16] researched on a model for predicting the performance of multilateral well, and as such, they developed a deliverability model. Firstly, a model that describes the performance of each lateral of the well was developed, coupling a model that describes inflow in a reservoir model with a model that described flow in wellbore to estimate the performance and volume of flow contributed by each lateral. The lateral model that was developed considered pressure drop. Their developed multilateral deliverability model could be used to estimate the performance of each lateral, the whole performance and the total pressure variation in the multilateral well.

Guo et al. [10] stated that although it has become common in the industry to drain a reservoir with a horizontal well and multilateral wells, it was observed that most of these wells do not produce at their expected production rate. This is because it is difficult to estimate the exact 'expected production rate' due to the fact that the production rate is estimated by models which stem from the assumption that the well was an infinite-conductive drain hole by considering the frictional effect of the long horizontal portion of the flow.

A semi-analytical model is reported by Tabatabaei and Ghalambor [8] for predicting the horizontal oil well performance. The model couples flow from a box-shaped drainage volume to flow in the wellbore. The horizontal wellbore flow description presented considers pressure drop due to friction, acceleration and fluid in flow effect. Their model easily adapted to predict productivity of multilateral wells by coupling the inflow performance of individual laterals with build-up section and the main vertical. The outcome of their study was more accurate than other previous experts as it shows the least percentage error derivation of 5% from the actual result obtained from gauge measurement. The recent study by Fadairo et al. reveals that all possible pressure restriction terms should be considered to combat the inaccuracy in results obtained using existing models in the literature [1–3, 17–19]. This chapter is an advancement on the Tabatabaei and Ghalambor model [8] by inclusion of pressure restriction due to accumulation in the governing flow equation for horizontal well. The output of this research shows that the disparity between the measured gauge value and previous work done is due to their failure to consider all possible pressure drops in long horizontal drain hole including pressure drop due to accumulation as the present study gives less than 1.2% error deviation from the actual value.

## 2. Theory

The numerical approach is more reliable for field application than the analytical approach. However, the numerical approach involves a systematic procedure and iterative nature which require longer computational times. It is more difficult to compute and access for day-to-day application in the industry. A basic and thorough semi-analytical approach has capacity to accurately predict the performance of a horizontal well. It is attractive and simpler to use as well as extensive and sufficiently exact for field applications if the governing fundamental flow equation is accurately modelled.

Generally, the existing models describing the performance of horizontal wells are divided into three classifications:

1. Analytical solutions.

2. Semi-analytical models.

3. Numerical models.

Semi-analytical coupling model gives an exhaustive and comprehensive estimate of productivity; this model is applicable do different reservoirs of varying conditions. Similarly, this model can be easily modified to predict the productivity of multilateral wells by coupling the inflow from all the different laterals with the total hydraulic build-up in the wellbore [8].

## 3. Model description

The horizontal and multilateral inflow model derived from the coupling of porous media inflow and horizontal drain hole inflow models have been reported by several experts in the literature. One of the earliest coupling models was developed by Dikken [4], and afterward, several others have been reported. The results obtained from previous models show large disparity between the actual and the calculated result for failure to consider all accessible pressure drop in the horizontal drain portion.

Consider fluid flow from the reservoir into the horizontal drain hole as shown in **Figure 1**. Assuming that the reservoir is assumed to be a constant pressure reservoir with the outer boundary responsible for keeping the pressure constant, and as such, the reservoir pressure is assumed to be the outer boundary pressure $P_e$. Flowing pressure along the horizontal well is not constant and hence does not only depend on pressure drop due to friction and acceleration as opined by Tabatabaei and Ghalambor [8] but also based on restriction due to accumulation. The general coupling inflow equation for the horizontal well system is expressed as

$$q_s(x) = J_s(x)[P_e - P_w(x)] \tag{1}$$

In this paper, the reservoir productivity index $J_s$ can be obtained using the Furui et al. [21] model while the flowing horizontal wellbore pressure can be obtained from the fundamental energy equation of flow in pipe as a function of space and time.

The overall flow rate of the horizontal well is gotten by the integration of Eq. (1) along the entire length of wellbore

$$Q = \int_0^L q_s(x)dx = \int_0^L J_s(x)[P_e - P_w(x)]dx \tag{2}$$

In solving Eq. (2), an analytical solution is more cumbersome because the pressure along the wellbore $P_w(x)$ and the specific productivity index $J_s(x)$ vary with the length of the well and several pressure dependent variables as function of time and space are involved, and hence, the coupling model is solved numerically.

To resolve Eq. (2), the lateral portion of the well is divided into a small number of segments; these segments are numbered from the toe to the heel as demonstrated in **Figure 2**. Therefore, the overall flow rate is an addition of the flow rates from the different segments.

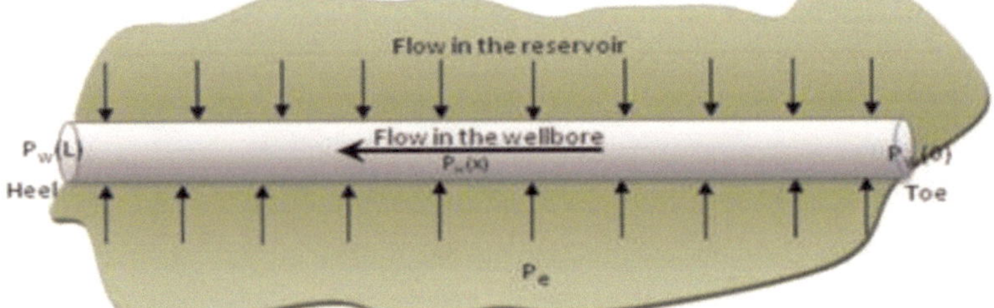


**Figure 1.**
*Coupled wellbore-flow and reservoir in-flow [20].*

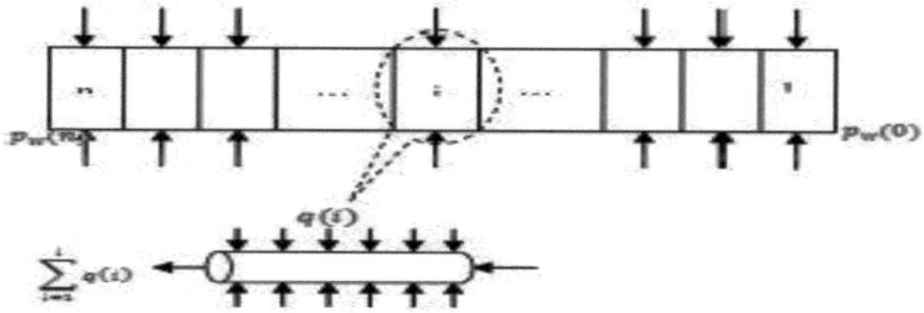

**Figure 2.**
*A diagram of a segmented lateral of the wellbore [8].*

$$= \sum_{i=1}^{i=n} q(i) \tag{3}$$

Accordingly, to determine the flow rate of every segment $q(i)$, the wellbore flow at a segment is coupled with the reservoir inflow throughout that segment

$$q(i) = J_s(i)\Delta x\left[P_e - \overline{P_w}(i)\right] \tag{4}$$

It is assumed that the length of the segment, $\Delta x$, is very small, and as such, the specific productivity index of the segment $J_s(i)$ does not vary along the segment, as such it is computed at the centre of every segment by using the model developed by Furui et al. [21]:

$$J_s(i) = \frac{7.08 \times 10^{-3} k}{\mu B\left[In\left(\frac{hl_{ani}}{r_w(l_{ani}+1)}\right) + \frac{\pi w}{2hl_{ani}} - 0.785 + S(i) + S_R\right]} \tag{5}$$

The partial-penetration skin factor $S_R$ is computed using the model created by Babu and Odeh [22]. Similarly, the reservoir anisotropy and the exposure time to drilling fluid (especially drilling mud) are assumed. Also, the elliptical-cone-shaped model is assumed for the distribution of formation damage factor $S(i)$ along the lateral of the well suggested by Frick and Economides [23].

$$S(i) = \left(\frac{k}{k_s} - 1\right) In\left[\left(\left[\frac{2a_{max}}{r_w[I_{ani} + 1]} - 1\right]\frac{x(i)}{L} + 1\right)\right] \tag{6}$$

To calculate the average pressure in the horizontal wellbore throughout the segment $\overline{P_w}(i)$, the following equation is used:

$$\overline{P_w}(i) = \frac{1}{2}[P_w - P_w(i-1)] \tag{7}$$

where

$$P_w(i) = P_w(i-1) - \Delta P_{fric}(i) - \Delta P_{acc}(i) - \Delta P_{acm}(i) \tag{8}$$

The two above equations are combined to give Eq. (9)

$$\overline{P_w}(i) = P_w(i-1) - \frac{1}{2}\left[\Delta P_{fric}(i) + \Delta P_{acc}(i) + \Delta P_{acm}(i)\right] \tag{9}$$

The pressure drop due to acceleration and friction along the wellbore was obtained using a flow model developed by Ouyang et al. [9] while the pressure drop due to accumulation was obtained using the concept reported by Fadairo et al. [1].

The pressure drop due to friction throughout every segment for both laminar flow regime and turbulent flow regime in oilfield units is determined by the equation as follows:

Pressure drop due to friction in the laminar flow regime:

$$\Delta P_{fric}(i) = C_1[q_t + q(i)]\left(1 + C_2[q_t + q(i)]^{0.6142}\right) \tag{10}$$

where

$$C_1 = \frac{8 \times 10^{-6}\mu\Delta x}{d^4} \tag{11}$$

$$C_2 = 5.08 \times 10^{-3}\left[\frac{\rho}{\mu x(i)}\right]^{0.6142} \tag{12}$$

and the $q_t$ is the axial-flow rate going into the segment, and this is shown as:

$$q_t = \sum_{i=1}^{i-1} q(i) \tag{13}$$

Pressure drop due to friction in the turbulent-flow regime:

$$P_{fric}(i) = \frac{C_3\left([q_t + q(i)]^2 - \left(C_4[q_t + q(i)]^{2.3978}\right)\right)}{\left[-4\log\left(C_5 - \frac{C_6}{[q_t+q(i)]} \times \log\left(C_7 + \frac{C_8}{[q_t+q(i)]^{0.8981}}\right)\right)\right]^2} \tag{14}$$

where

$$C_3 = \frac{7.46 \times 10^{-7}\mu\Delta x}{d^5} \tag{15}$$

$$C_4 = 3.83 \times 10^{-3}\left[\frac{\rho}{\mu x(i)}\right]^{0.3978} \tag{16}$$

$$C_5 = \frac{\varepsilon}{3.7065} \tag{17}$$

$$C_6 = \frac{3.385\mu d}{\rho} \tag{18}$$

$$C_7 = \frac{\varepsilon^{1.1098}}{2.8257} \tag{19}$$

$$C_8 = 4.09\left(\frac{\mu d}{\rho}\right)^{0.8981} \tag{20}$$

Pressure drop in acceleration throughout every segment is determined by:

$$\Delta P_{acc}(i) = \frac{7 \times 10^{-9}\rho}{g_c d^4}\left[q^2(i) + 2q_t q(i)\right] \tag{21}$$

Pressure drop due to accumulation is to be determined by:

$$\Delta P_{acm}(i) = \frac{4.1667 \times 10^{-5}}{g_c d^2 t}[q(i) + q_t] \tag{22}$$

An iterative method is used to solve Eq. (4) as the pressure drop is associated with the production rate. The procedure for calculating the production rate of each segment $q(i)$ and the overall cumulative production rate $Q$ is as follows:

1. A pressure for the wellbore at the toe is assumed, $P_w(0)$.

2. The portrayed reservoir/wellbore-coupling model is used to determine flow rate, (1), and pressure drop over Segment 1. $[\Delta P_{fric}(i) + \Delta P_{acc}(i) + \Delta P_{acm}(i)]$.

3. Equation (8) is used to compute the pressure at the end of segment 1, $P_w(1)$, Eq. (8):

$$P_w(i) = P_w(i-1) - \Delta P_{fric}(i) - \Delta P_{acc}(i) - \Delta P_{acm}(i)$$

4. Steps 2 and 3 will be repeated, advancing in the direction of the heel to ascertain the flow rate in every segment, $q(i)$, and then the pressure at the end of every segment, $P_w(i)$, can be calculated.

5. The flowing bottom hole pressure, $P_{wf}$, and the pressure that has been calculated at the heel, $P_w(n)$, with Eq. (8) are compared and the pressure at the end of each segment can be calculated; as such, the flow rate in each segment can be determined with Eq. (3). Thereafter, the flow rate from each section will be summed up to give the total flow rate.

If the condition in Eq. (23) is not true, then another value is assumed for the pressure at the toe and the procedure from step 2 to step 5 is repeated until the condition in Eq. (23) is true

$$|P_{wf} - P_w(n)| \le e \tag{23}$$

Here, the estimation of e relies upon the degree of accuracy required in expectation of well efficiency.

## 4. Multilateral-well deliverability model

The concept in the currently developed model for horizontal productivity can be adapted to evaluate flow in a multilateral well by commingling flow from different lateral or horizontal portions into a main wellbore. **Figure 3** shows a multilateral well with three lateral wells. The pressure that is known at the beginning is the wellhead pressure, and every other component of pressure in the well system are and must be resolved. The following process is to predict the pressure drop behaviour along each lateral, and also the pressure drop behaviour in the main borehole with the corresponding production performance of each lateral and overall production rate can also be predicted. A pressure for the wellbore at the toe for the first lateral is assumed, $P_w(0, 1)$. The coupling equation from the previous section is used

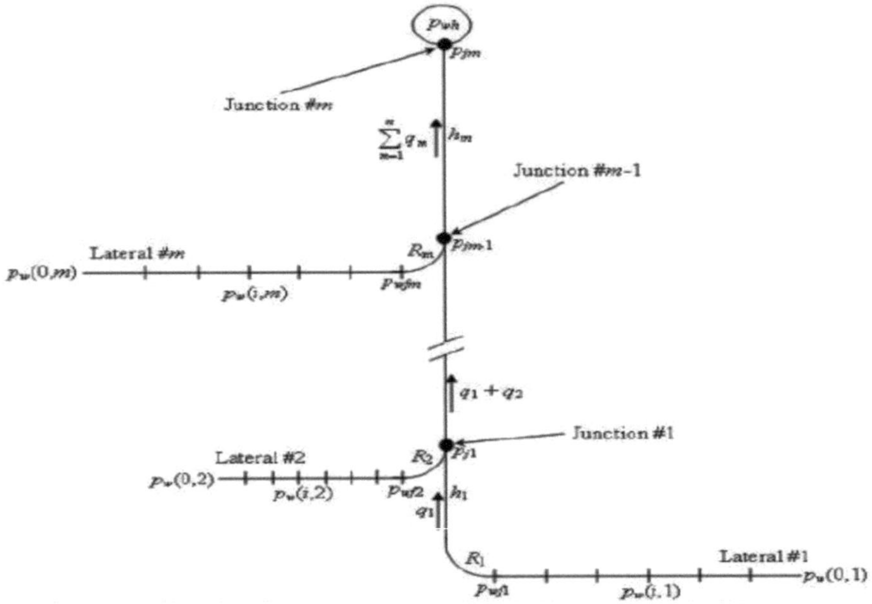


**Figure 3.**
*A diagram of a multilateral well [8].*

to determine the productivity of lateral 1 and the pressure performance along the first lateral to ascertain the pressure of the heel of lateral 1 $P_{wf1}$.

1. The pressure at junction 1 is obtained with the following equations:

$$P_{j1} = P_{wf1} - \Delta P_{gravity|R1\&h1} - \Delta P_{friction|R1\&h1} \tag{24}$$

where

$$\Delta P_{gravity|R1\&h1} = \frac{\rho_1(R_1 + h_1)}{144} \tag{25}$$

and

$$\Delta P_{friction|R1\&h1} = \frac{f_f \rho_1 v_{t1}^2 \left(\frac{\pi}{2} R_1 + h_1\right)}{6 g_c d_t} \tag{26}$$

2. A pressure for the wellbore at the toe for the second lateral is assumed, $P_w(0,2)$. The coupling equation from the previous section is used to determine the productivity of lateral 2 and the pressure performance along the second lateral to ascertain the pressure of the heel of lateral 1 $P_{wf2}$.

3. A new pressure at junction 1 is $P_{j1}$ and is calculated with Eq. (27):

$$P_{j1} = P_{wf2} - \frac{\rho_2 R_2}{144} - \frac{f_f \rho_2 v_{R2}^2 \pi R_2}{12 g_c d_t} \tag{27}$$

4. Make a comparison of $P_{j1}$ gotten from step 2 to that from step 4, and steps 3 and 4 will be repeated until the two values of $P_{j1}$are similar and as such the production performance from lateral 2 and the pressure performance from lateral 2 are known.

5. Sum the flow rate of lateral 1 and lateral 2 to determine the total production rate between junction 1 and 2.

6. The pressure at the second junction $P_{j2}$ can be calculated using Eq. (28), and similarly, all the subsequent pressures at the different junctions, using the same equation:

$$P_{jm} = P_{jm-i} - \frac{\rho_{avg} h_m}{144} - \frac{f_f \rho_{avg} v_{tm}^2 h_m}{6 g_c d_t} \tag{28}$$

where

$$\rho_{avg} = \frac{\sum_{m=1}^{m} \rho_m q_m}{\sum_{m=1}^{m} q_m} \tag{29}$$

7. Steps 3 and 8 will be repeated, moving upwards on the main wellbore to determine the production performance of the other laterals that might be present, and furthermore, the pressure present at every junction.

8. A comparison is made between the pressure calculated at the junction m, $P_{jm}$, to the pressure at the wellhead. $P_{wh}$. The overall flowrate of the well system is gotten from adding up the production rate from each lateral.

$$\left| P_{wh} - P_{jm} \right| \le e \tag{30}$$

In the event that Eq. (30) does not hold, another pressure value must be assumed at the toe for the wellbore of the first lateral, $P_w(0, 1)$, and the entire methodology ought to be repeated.

## 5. Results and discussion

To validate the current model on the productivity of horizontal and multilateral wells, the field data from a horizontal well in Australia as reported by Tabatabaei and Ghalambor [8] and presented in **Table 1** was employed. Additionally, in this section, interactive plots of the estimated well pressure, production profile and

| Parameters | Value |
|---|---|
| Length of the reservoir | 2438 ft |
| Width of the reservoir | 600 ft |
| Height of the reservoir | 131.2 ft |
| Lateral length | 2438 ft |
| Radius of wellbore | 0.354 |
| Length from middle to the boundary | 1219 ft |
| Effective wellbore diameter | 5.5 in |
| Roughness of the wellbore | 0.1in |
| Vertical permeability | 345md |
| Horizontal permeability | 850md |
| Formation damage permeability | 100md |

| Parameters | Value |
|---|---|
| Skin factor due to invasion | 2 |
| Skin factors due to other factors | 5 |
| Pressure of the reservoir | 932.5 psi |
| Pressure of wellbore at heel | 925 psi |
| Viscosity of oil | 0.5cp |
| Oil formation volume factor | 1.058rb/Stb |
| Density of oil | 55.97lbm/ft$^2$ |

**Table 1.**
*Field parameters [11].*

total pressure using current study compared with other existing models in literature were presented. **Table 2** presents the reservoir and well properties of the multilateral wells that were used as an input for predicting multilateral flow performance. **Table 3** shows the comparison of production rate results obtained from the current model and other existing models in the literature using data in **Table 2** as an input. Performance of dual-lateral well with variation in wellbore pressure at different segments was equally evaluated as shown in **Table 4**. To analyse the time of well's stability, that is, how long it would take for the well to experience stabilised flow, plots of pressure and productivity at the heel and toe of the well were generated.

| Parameters | Lateral no. 1 | Lateral no. 2 |
|---|---|---|
| Length of the reservoir | 2500 ft | 2000 ft |
| Width of the reservoir | 750 ft | 500 ft |
| Height of the reservoir | 75 ft | 50 ft |
| Lateral length | 2400 ft | 1500 ft |
| Radius of wellbore | 0.325 ft | 0.325 |
| Length from middle to the boundary | 1215 ft | 900 |
| Effective wellbore diameter | 4.5 in. | 4.5 in |
| Roughness of the wellbore | 0.0024in | 0.0024in |
| Radius of build-up section | 50 ft | 30 ft |
| Distance to upper junction | 500 ft | 2500 ft |
| Vertical permeability | 25md | 50md |
| Horizontal permeability | 100md | 150md |
| Formation damage permeability | 10md | 25md |
| Skin factor due to invasion | 3 | 2 |
| Skin factors due to other factors | 5 | 5 |
| Pressure of the reservoir | 2250 psi | 2000 psi |
| Viscosity of oil | 0.5cp | 0.6cp |
| Oil formation volume factor | 1.2rb/stb | 1.25rb/stb |
| Density of oil | 56lbm/ft$^3$ | 58lbm/ft$^3$ |

**Table 2.**
*Reservoir and well properties of each lateral [8].*

| Model | Production rate (stb/d) | Error (%) |
|---|---|---|
| Actual | **5660** | 0 |
| Economides et al. [24] | 8324 | 47 |
| Furui et al. [21] | 8405 | 48 |
| Guo et al. [10] | 5152 | 9 |
| Tabatabaei et al. [8] | 5939 | 5 |
| The current model | **5593** | **1.19** |

**Table 3.**
*Comparison of the productivity from different models.*

| Results from each lateral | Current model | | Tabatabaei and Ghalambor [8] | |
|---|---|---|---|---|
| | Lateral 1 | Lateral 2 | Lateral 1 | Lateral 2 |
| Production from each lateral (STD/D | 2470.38 | 14260.47 | 24,994 | 14,274 |
| Pressure of wellbore at toe (psi) | 1900.8 | 1646.56 | 1899 | 1645 |
| Pressure of wellbore at heel (psi) | 1848.7 | 1634.55 | 1874 | 1633 |
| Pressure of wellbore at junction | 1620.4 | 500.2 | 1620 | 500 |

**Table 4.**
*Results of production prediction from each lateral and pressure at each junction.*

## 6. Model validation and comparison

The productivity prediction model for horizontal wells presented in this paper is verified at field scale using the case study presented by Chauvel et al. [11] as discussed in Tabatabaei and Ghalambor [8]. The horizontal well exhibits an 8.5-in open hole completed using a 5.5-in pre-packed screen opened laterally along the well length. The well trajectory is reported almost perfectly horizontal in 131.2 ft oil pay zone thickness and overall vertical depth of less than 6 in.

Production data indicated a liquid flow rate of 5677 BOPD, which corresponds to surface measured production rate with little free gas as the well was producing some psi below the bubble point. As reported in Tabatabaei and Ghalambor [8], some important parameters such as reservoir permeability, skin factor and boundaries were not reported. Therefore, as discussed by Tabatabaei and Ghalambor [8], these parameters were estimated (for the purpose of model validation) by matching the wellbore pressure profile calculated by this current model to measured pressure data.

**Figure 4** presents the predicted pressure profile using the current study and the predicted pressure profile using Tabatabaei and Ghalambor [8] along the wellbore adopting the optimum segment number idea of 15. As it can be observed, the inclusion of the accumulation term into the current model as an improvement in Tabatabaei model resulted in a lesser pressure data at the heel and a higher pressure at the toe. The same trend was observed in the plot of specific inflow at each segment within the well as shown in **Figure 5**.

Using the parameters in **Table 1**, the current model can be used to predict productivity of horizontal well, and the results are compared with the actual production rate to ascertain the level of accuracy and precision of the current study as reported in **Figure 6**. **Table 3** illustrates the results of this analysis; the current study that incorporates the accumulation term in the pressure drop equation

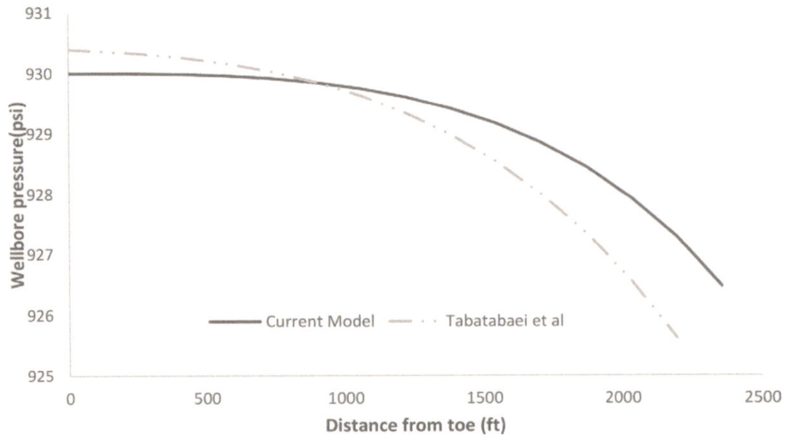


**Figure 4.**
*Pressure performance of the well against distance from the toe.*

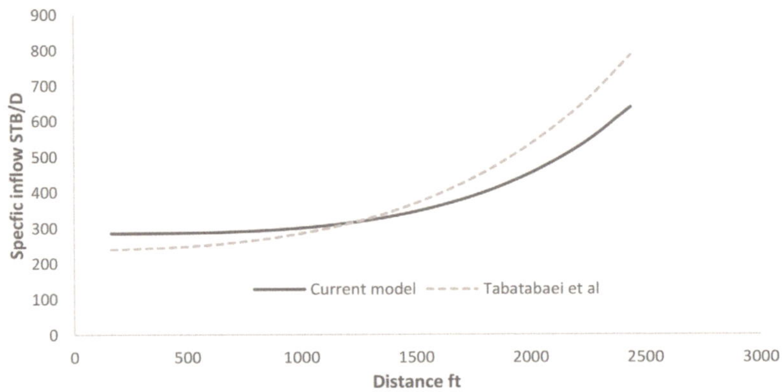


**Figure 5.**
*Specific inflow at each point in the well.*

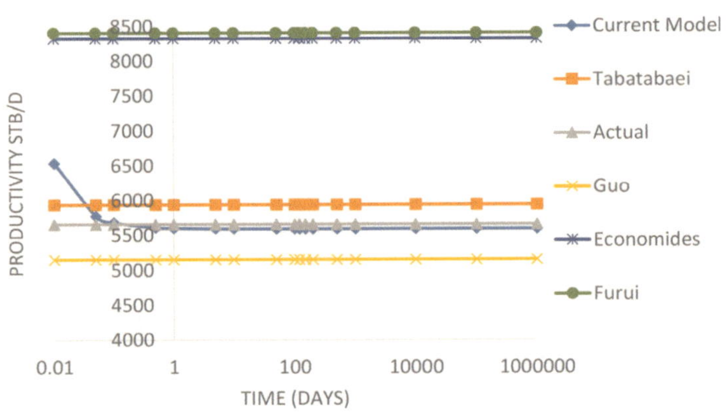


**Figure 6.**
*Productivity over time.*

predicted a production rate of 5593 STB/D compared to an actual production rate of 5660 STB/D reported by Chauvel et al. The close results validate the inclusion of the accumulation term in the coupling model and hence show why the model prediction gave the lowest percentage error 1.19%. Other models used for comparison were Economides et al. [24], Furui et al. [21], Guo et al. [7, 10] and Tabatabaei and

Ghalambor [8], as reported by Tabatabaei and Ghalambor [8], Economides and Furui assumed infinite-conductivity-drain hole, Guo included pressure drop along the wellbore in his development, while Tabatabaei and Ghalambor incorporated the acceleration term into the pressure drop equation. Using the same parameters as in **Table 1**, Economides et al. [24] and Furui et al. [21] overestimated the production rate (8324 STB/D and 8405 STB/D respectively) compared to the actual production rate which is evident in the recorded high percentage error. This was explained to be due to the omission of pressure drop along the wellbore in their model development. Guo et al. [7, 10] and Tabatabaei and Ghalambor [8] predictions are close to the actual production rate but not as accurate as the current study. Guo et al. model underestimated the well's productivity by around 9% because flow restriction is not only due to friction but all other pressure drops in horizontal wellbore such as pressure drop due to accumulation and in flow effect. The comparison also shows that the model by Tabatabaei and Ghalambor overestimated the well's productivity by approximately 5% for their failure to consider possible pressure drop due to accumulation in the wellbore. Therefore, the current study justified the inclusion of the accumulation term in the governing inflow equation for coupling model of reservoir-horizontal wellbore development.

The productivity of horizontal well depends on the difference between the reservoir pressure and the wellbore pressure at any point along the wellbore. Estimating the lateral productivity of horizontal well necessitates predicting the pressure profile and distribution along the wellbore. **Figures 6** and 7 respectively illustrate the well pressure and productivity distribution with time using the current model and Tabatabaei model. The current model exhibits both early time unstabilised flow and later time stabilised flow characteristics. The stabilised flow period accurately matched the actual productivity recorded on field; this analysis further justifies the introduction of the accumulation term in the current study and validates the accurate predictive power of the current model in terms of horizontal well's productivity prediction. Using the current model, stabilised flow period started at around 120 days and productivity of the well peaked at about 5592.8 STB/D.

Generally, the productivity of horizontal wells as a function of the wellbore length depends on the reservoir and wellbore properties. **Figure 8** presents the effect of horizontal well length on pressure drop and in turn productivity for predictions of both the current study and that of Tabatabaei [8]. It can be observed from the plot that productivity increases as the well length increases. Using the reservoir parameters and well completion information presented by Tabatabaei and Ghalambor [8] as an input in the current model and the existing model in the literature [8]. Multi-lateral well performance prediction by this model is illustrated

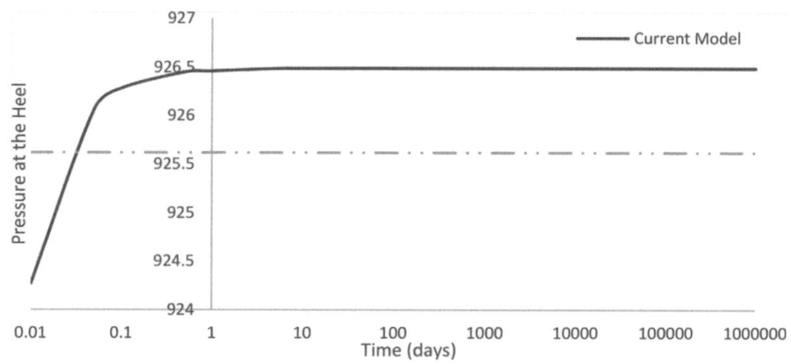


**Figure 7.**
*Pressure variation at the heel with time.*

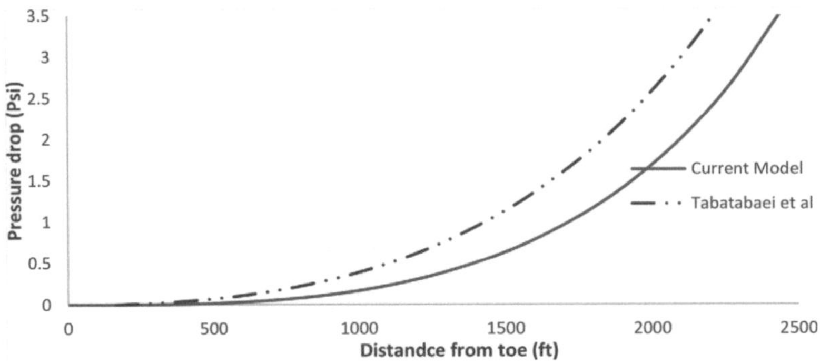


**Figure 8.**
*Effect of horizontal well length on productivity.*

and the outcomes are compared with those of Tabatabaei and Ghalambor [8] in **Table 4**. Total production and production from each lateral were calculated using the current model at different wellhead pressures. The result shows a non-linear increase in productivity as wellhead pressure increases.

## 7. Conclusion

The modified semi-analytical model was developed to predict the production performance of horizontal wells. A coupling of the inflow in the reservoir and flow in the wellbore was used for the development of the model.

Conclusions made from this study are as follows:

1. Disregarding the pressure drop in the wellbore will lead to an overestimation of the production rate of the well. Also, we see that not considering the inflow effect of the fluid will result in an underestimation of the production from the well.

2. Using the information gotten from field, it was shown that this model is more precise on the account that it gives a more practical representation of the flow in the wellbore and inflow to the reservoir, as this model is compared with the pre-existing models.

3. The model is simplified to be user-friendly as well as very efficient and sufficiently accurate for field applications. It can be used in reservoirs of varying conditions. We also see that the model is applicable for predicting the productivity of a well in a heterogeneous reservoir.

4. It can also be easily adapted to predict the productivity of multilateral well by incorporating the production performance of each lateral individually, with the well hydraulics of each of the build-up sections between the laterals and also the well hydraulics in the main wellbore.

5. Effects of friction, acceleration and accumulation, which lead to the pressure drop with increasing well lengths, should take into account avoiding an overestimation of the well's productivity.

## Nomenclature

| | |
|---|---|
| $q_s(x)$ | in flow in the well per unit length of the wellbore |
| $J_s(x)$ | specific productivity index |
| $J_s(i)$ | specific productivity index of segment number i |
| $\overline{P_w}(i)$ | average wellbore pressure at this segment |
| $\Delta x$ | length of the segment |
| $S_R$ | partial-penetration skin factor |
| $S(i)$ | formation damage skin factor |
| $k$ | effective permeability of reservoir |
| $x(i)$ | distance between the centre of the segment I and the toe |
| m | number of junctions |

## Acknowledgements

We wish to thank all members of staff of the Department of Petroleum Engineering, Covenant University, Nigeria, for their support in carrying out this research work. We also acknowledge the technical support of Environmental and Energy Research Team, FatherHeroz Forte Technology Nigeria Limited for their commitment.

## Author details

Adesina Fadairo[1*], Gbadegesin Adeyemi[1], Temitope Ogunkunle[1], Ayotomiwa Evbogame[1] and Adedapo Adesina[2]

1 Department of Petroleum Engineering, Covenant University, Ota, Nigeria

2 M-I Swaco Schlumberger Oil Services Company, Equatorial Guinea

*Address all correspondence to: adesinafadairo@yahoo.com

## References

[1] Adesina F, Adebowale O, Gbadegesin A. A modified predictive model for estimating gas flow rate in horizontal drain hole. Petroleum. 2018. DOI: 10.1016/j.petlm.2018.05.001

[2] Adesina F, Paul A, Preye O, Adebowale O. An improved model for estimating productivity of horizontal drain hole. SPE Paper 184264, SPE Nigerian Annual International Conference and Exhibition, Lagos. 2016. pp. 2-4. DOI: 10.2118/184264-MS

[3] Adesina F, Churchill A, Falode O. Modelling productivity index for long horizontal well. Journal of Energy Resources Technology (JERT-09-1071). 2011;**133**. DOI: 10.1115/1.4004887

[4] Dikken B. Pressure drop in horizontal wells and its effect on production performance. Journal of Petroleum Technology. 1990;**42**. DOI: 10.2118/19824-PA. SPE-19824-PA

[5] Landman MJ. Analytical modeling of selectively perforated horizontal wells. Journal of Petroleum Science and Engineering. 1994;**10**:179-188. DOI: 10.1016/0920-4105(94)90079-5

[6] Guo BL. A rigorous composite IPR model for multilateral wells. SPE-ATCE. San Antonio,Texas: SPE 100923;2006. DOI: 10.2118/100923-MS

[7] Guo B, Ali C, Zhang L. A simple analytical model for predicting gas productivity of horizontal drain holes, with high friction. SPE Western Regional Meeting. Los Angeles, CA: SPE 108390;2007. DOI: 10.2118/108390-MS

[8] Tabatabaei M, Ghalambor A. A new method to predict performance of horizontal and multilateral wells. SPE 141164 2011;13

[9] Ouyang LB, Arbabi S, Aziz K. A simplified approach to couple wellbore flow and reservoir inflow for arbitary well configurations. SPE-48936-MS; 1998

[10] Guo B, Zhou J, Liu Y, Ghalambor A. A rigourous analytical model for fluid flow in drain hole for finite conductivity applied to horizontal and multilateral wells. SPE 106947;2007

[11] Chauvel Y, Anderson J, Aldred R, dePledge L. Horizontal well production logging in Australia. SPE 28757 1994;18

[12] Novy R. Pressure Drop in Horizontal wells: When Can they be ignored? 1995;6

[13] Ozkan E, Haciislamoglu M. Effect of conductivity on horizontal-well pressure-behaviour. SPE-24683-PA; 1995;10

[14] Penmatcha VR, Arbabi S, Aziz K. Effects of Pressure Drop in Horizontal Wells and Optimum Well Length. Presented at 1997 SPE Production Operations Symposium, Oklahoma City, Oklahoma, March 9-11, SPE Paper 37494;1997. DOI: 10.2118/37494-MS

[15] Aziz K, Penmatcha VR. Comprehensive reservoir/wellbore model for horizontal wells. SPE-57194-PA;1999

[16] Chen W, Hill AD, Zhu D. A comprehensive model of multilateral well deliverability. SPE-64751;2000

[17] Adesina F, Gbadegesin A, Temitope O, Adebowale O. A realistic model for estimating productivity index of vertical well using wellhead data. SPE Paper 193506, SPE Nigerian Annual International Conference and Exhibition, Lagos; 6-8 August;2018

[18] Adesina F, Falode O, Nwosu C. A new model for predicting liquid loading in a gas well. Journal of Natural Gas

Science and Engineering. 2015; **26**(2015):1530-1541. DOI: 10.1016/j.jngse.2014.09.003

[19] Adesina F, Olusegun A, Temitope OAO. Modeling the effect of sand particles on pressure transverse in flowing solid-gas well. World Journal of Engineering. 2017;**14**(2017):47-54. DOI: 10.1108/WJE-11-2016-0130

[20] Ulaeto UW, Onyekonwu M, Ikiensikimama S. Improved deliverability model for horizontal and multilateral wells. SPE-172433-MS;2014

[21] Furui K, Zhu D, Hill A. A rigorous formation damage skin factor and reservoir inflow model for a horizontal well. SPERE, 151 SPE-74698-MS;2003. DOI: 10.2118/74698-MS

[22] Babu DK, Odeh AS. Productivity of a Horizontal Well. SPE-18298-PA 1989;29

[23] Frick TP, Economides MJ. Horizontal well damage characterization and removal. SPE 21795, 1993;8

[24] Economides MJ, Deimbacher FX, Brand CW, Heinemann ZE. Comprehensive simulation of horizontal-well performance. 1991;8

Chapter 5

# Gases Reservoirs Fluid Phase Behavior

*Eman Mohamed Mansour, Mohamed El Aily*
*and Saad Eldin Mohamed Desouky*

## Abstract

This chapter discusses the fundamentals of the phase behavior of hydrocarbon fluids. Real reservoir fluids contain many more than two, three, or four components; therefore, phase-composition data can no longer be represented with two, three or four coordinates. Instead, phase diagrams that give more limited information are used. The behavior of reservoir of a reservoir fluid during producing is determined by the shape of its phase diagram and the position of its critical point. Many of producing characteristic of each type of fluid will be discussed. Ensuing chapters will address the physical properties of these three natural gas reservoir fluids, with emphasis on retrograde gas condensate gas, dry gas, and wet gas.

**Keywords:** phase behavior, reservoirs fluid, physical properties

## 1. Introduction

Petroleum reservoirs are mixtures of hydrocarbon organic compounds that may be in the liquid state or in a gaseous state or in combinations of gas and liquid as will describe in this chapter [1]. The most important part in petroleum engineering for production and reservoir engineers is studying hydrocarbon phase behavior of reservoirs and characteristics of it early in the life of reservoir to suggest maximize development in the future [2]. Petroleum reservoirs can be classified into gas reservoirs, oil reservoirs, and this classification according to phase behavior diagram. This category of natural gas reservoirs is a unique type of hydrocarbon system because it has special thermodynamic behavior of the gas reservoir fluid that controlling in development [3]. To predict the original of natural gas in place, we use many equations as material balance equations [4]. This chapter describes the gas reservoirs principle only and we will continue description oil reservoirs in another chapter.

## 2. Classification of gas reservoirs fluids

In general, reservoirs temperature is more than the hydrocarbon fluid critical temperature, the reservoirs are considered as a natural gas reservoir [5]. There are three types of gas petroleum reservoirs subdivided into retrograde gas, wet gas, and dry gas [3]. All this gas reservoir fluid type can be determined by experimentally working and by the stock-tank liquid gravity (API), the color of liquid, heptane plus and producing a gas-oil ratio. These differences in phase behavior lead to different physical properties



IntechOpen

for each reservoir. This classification according to initial formation temperature and pressure, production surface temperature and pressure and composition of the reservoir fluid. In addition, the classification of hydrocarbon fluids can be by the composition analysis of the fluid mixture, where it is one of strongest effect on the fluid characteristics as shown in the ternary diagram (**Figure 1**) [6].

The diagram conditions under which these phases expressed is a pressure-temperature diagram or phase diagrams, where these diagrams are a different multicomponent system with a different phase diagram [7, 8]. The gases phase's diagrams are used to define the phase behavior and natural of these three types of hydrocarbon systems. To understand any gases phase's diagrams, it is necessary to define these key points on these diagrams [9]:

- **Hydrocarbon phase envelope:** it is region enclosed by the dew-point curve, where gas and liquid coexist in equilibrium phase. In addition, it can be called by two-phase region.
- **Dew-point pressure:** it is pressure at which separating the vapor-one phase region from the two-phase region.
- **Critical point:** it is pressure $P_c$ and temperature $T_c$ of the mixture hydrocarbon at which liquid and gas phase's properties are equal.
- **Cricondenbar (Pcb):** it is a maximum pressure above which no gas can be formed regardless of temperature.
- **Cricondentherm (Act):** it is a maximum temperature above which no liquid can be formed regardless of pressure.
- **Quality lines:** it is dashed lines inside the phase diagram that define the temperature and pressure for equal volumes of liquids [10].

Depending on reservoir conditions, natural gases reservoirs fluids can be classified into:

- Retrograde gas-condensate
- Wet gas
- Dry gas

### 2.1 Retrograde gas-condensate reservoirs

The retrograde gas-condensate reservoir is also called retrograde condensate gases, condensate, retrograde gas and gas condensates. In this type of natural gas, reservoir prefer called gas-condensate and not condensate only because this reservoir exhibits retrograde behavior [11]. In case of the reservoir temperature more than a critical temperature and less than a critical temperature, the reservoir is classified as a retrograde gas-condensate reservoir as shown in **Figure 1**. As a result of the critical point of the retrograde gas phase is further down the left side of the envelope as shown in **Figure 2**, heavy hydrocarbons will be fewer as compared with oils [12].

In the bagging of the reservoir, the hydrocarbon system will be totally one phase gas (i.e., vapor phase) because the reservoir pressure is above the dew-point pressure. As the reservoir pressure decrease from the initial formation pressure through the production until dew-point pressure, where the liquid starts to condense from the gas

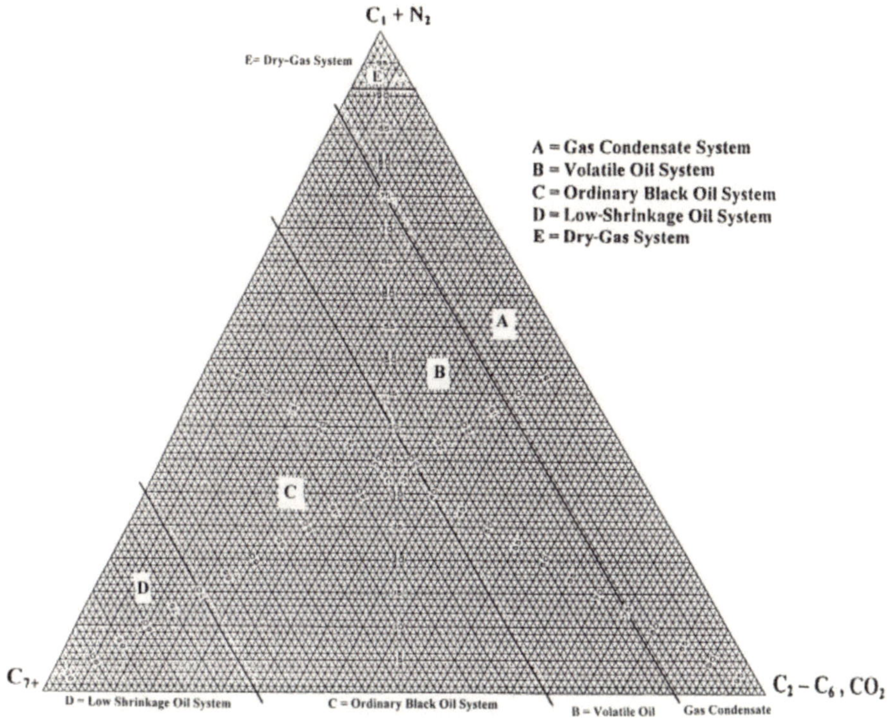


**Figure 1.**
*Compositions of various reservoir fluid types.*

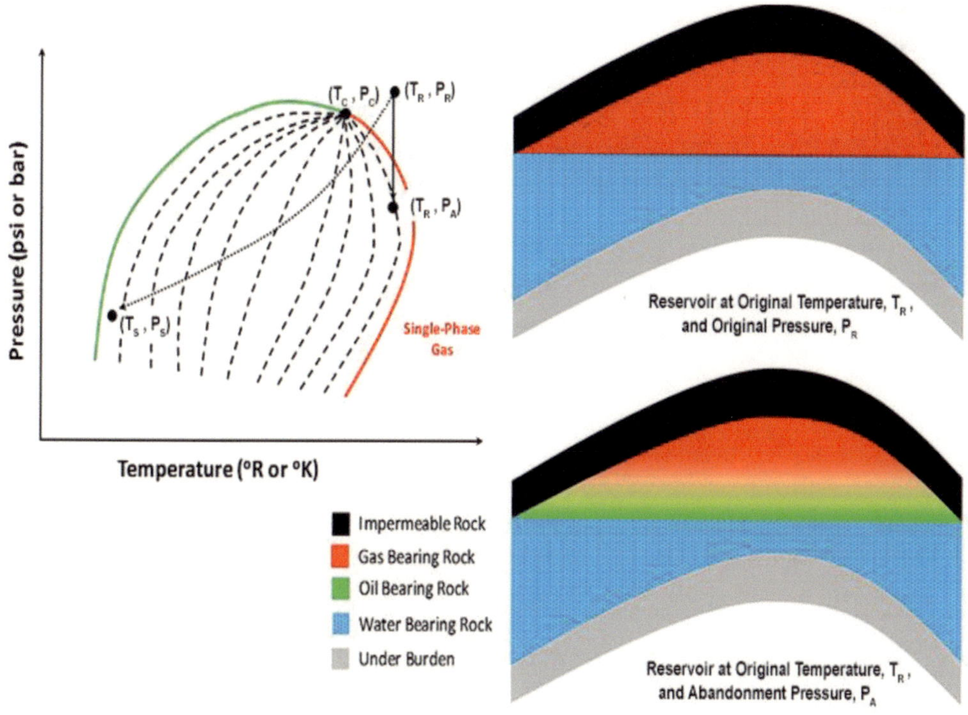


**Figure 2.**
*Retrograde gas-condensate reservoir phase diagram.*

| Reservoir information | Sample 1 | Sample 2 | Sample 3 |
|---|---|---|---|
| Reservoir pressure, psi | 6740 | 6243 | 5876 |
| Reservoir temperature, °F | 321 | 304 | 220 |
| $C_7^+$, Mole % | 6.3511 | 0.4408 | 1.4177 |
| Average mole weight | 30.38 | 23.78 | 22.11 |
| Dew point pressure, psi | 4433 | 5030 | 4854 |
| GOR, STB/SCF | 9088.49 | 26766.81 | 36155.894 |
| API | 49.30 | 51.55 | 60.64 |
| **Compositional analysis of reservoir fluid to C36+** | | | |
| **Component** | **Mole %** | **Mole %** | **Mole %** |
| Hydrogen | 0.0000 | 0.0000 | 0.0000 |
| Hydrogen sulfide | 0.0000 | 0.0000 | 0.0000 |
| Carbon dioxide | 7.5475 | 5.4118 | 0.4068 |
| Nitrogen | 0.9154 | 0.1125 | 0.0262 |
| Methane | 73.5809 | 76.2140 | 83.2475 |
| Ethane | 5.4077 | 9.3173 | 6.9531 |
| Propane | 2.5521 | 3.9001 | 4.0096 |
| i-Butane | 1.0076 | 0.7866 | 1.1481 |
| n-Butane | 1.0066 | 0.8496 | 0.9764 |
| Neo-Pentane | 0.0000 | 0.0000 | 0.0000 |
| i-Pentane | 0.7871 | 0.4966 | 0.5337 |
| n-Pentane | 0.1538 | 0.2918 | 0.5244 |
| Hexanes | 0.5606 | 0.5099 | 0.4908 |
| M-C-Pentane | 0.0397 | 0.0237 | 0.0846 |
| Benzene | 0.0372 | 0.0234 | 0.0438 |
| Cyclohexane | 0.0528 | 0.2295 | 0.1374 |
| Heptanes | 0.4160 | 0.4434 | 0.2230 |
| M-C-Hexane | 0.0226 | 0.0391 | 0.0773 |
| Toluene | 0.0381 | 0.1241 | 0.0682 |
| Octanes | 0.3214 | 0.1875 | 0.1866 |
| E-Benzene | 0.3010 | 0.1540 | 0.0071 |
| M/P-Xylene | 0.0273 | 0.0087 | 0.0600 |
| O-Xylene | 0.1804 | 0.0219 | 0.0229 |
| Nonanes | 0.8192 | 0.2011 | 0.1285 |
| 1,2,4-TMB | 0.0486 | 0.0097 | 0.0120 |
| Decanes | 1.0640 | 0.2024 | 0.1075 |
| Undecanes | 0.9636 | 0.1369 | 0.0992 |
| Dodecanes | 0.5167 | 0.0580 | 0.0791 |
| Tridecanes | 0.3672 | 0.0440 | 0.0655 |
| Tetradecanes | 0.2981 | 0.0438 | 0.0518 |

| Component | Mole % | Mole % | Mole % |
|---|---|---|---|
| Pentadecanes | 0.2494 | 0.0303 | 0.0466 |
| Hexadecanes | 0.2013 | 0.0267 | 0.0402 |
| Heptadecanes | 0.1556 | 0.0166 | 0.0351 |
| Octadecanes | 0.0973 | 0.0116 | 0.0217 |
| Nonadecanes | 0.0704 | 0.0185 | 0.0175 |
| Eicosanes | 0.0632 | 0.0072 | 0.0172 |
| Heneicosanes | 0.0387 | 0.0078 | 0.0098 |
| Docosanes | 0.0264 | 0.0063 | 0.0096 |
| Tricosanes | 0.0176 | 0.0097 | 0.0088 |
| Tetracosanes | 0.0126 | 0.0097 | 0.0073 |
| Pentacosanes | 0.0087 | 0.0033 | 0.0045 |
| Hexacosanes | 0.0073 | 0.0029 | 0.0034 |
| Heptacosanes | 0.0050 | 0.0020 | 0.0023 |
| Octacosanes | 0.0040 | 0.0016 | 0.0018 |
| Nonacosanes | 0.0025 | 0.0010 | 0.0013 |
| Triacontanes | 0.0036 | 0.0017 | 0.0008 |
| Hentriacontanes | 0.0019 | 0.0007 | 0.0006 |
| Dotriacontanes | 0.0006 | 0.0003 | 0.0003 |
| Tritriacontanes | 0.0004 | 0.0002 | 0.0002 |
| Tetratriacontanes | 0.0001 | 0.0001 | 0.0002 |
| Pentatriacontanes | 0.0003 | 0.0001 | 0.0001 |
| Hexatriacontanes plus | 0.0001 | 0.0001 | 0.0001 |
| Total | 100.00 | 100.00 | 100.00 |

**Table 1.**
*Examples of retrograde gas-condensate reservoirs.*

phase to form a free liquid in the reservoir as a result of molecules attraction between light and heavy components move further apart [13]. The condensate liquid still is inside the reservoir and cannot be produced from it. The condensate liquid volume not more than 15–19% of the pore volume, so this liquid still be inside the reservoir and cannot be produced as it is not large volume enough to flow. All of this indicates by reservoir pressure path as shown in the retrograde gas-condensate figure [14].

Physical characteristics identification:

- **Gas-oil ratios (GOR):** common gas-oil ratios between 8000 and 70,000 SCF/STB. But the lower gas-oil ratio is approximately 3300 SCF/STB and the upper limit is over 150,000 SCF/STB. In case of low gas-oil ratio condense the liquid may be reached to 35% or more. With time, the gas-oil ratio of condensate reservoir increases due to heavy components loss.

- **Stock-tank gravity (API):** is usually above 40° API stock-tank and increase as formation pressure decrease below dew point pressure.

- **Heptane's plus fraction:** is less than 12.5-Mole% by laboratory analysis. But in case heptane plus fraction is less than one percent, the retrograde liquid volume is small so it is negligible.

- **Color:** may be slightly colored, orange, brown, greenish and water-white, so color is not depended on indicator if this reservoir gas condensate or oil.

**Table 1** shows data of reservoir information and compositional analysis of reservoir fluid for three different examples of retrograde gas-condensate reservoirs.

### 2.2 Wet gas reservoirs

It is the second type of natural gas reservoir fluid. In this type, reservoir temperature exceeds hydrocarbon system cricondentherm, so the reservoir fluid always remains in the gas phase as the reservoir pressure decrease. No condensate liquid is formed in the formation as a result of the pressure path does not inside the phase envelope as shown in a wet gas phase diagram (**Figure 3**) [15]. Some of the liquid is formed at the surface due to separator conditions (separator pressure and temperature) still inside the phase envelope and is called condensate. The expression of "wet gas" does not mean that the gas is wet with water but means condensation that occurs at the surface [16].

Physical characteristics identification:

- **Gas-oil ratios (GOR):** is very high producing gas-oil ratios reached from 60,000 to 100,000 SCF/STB. During wet gas reservoir life, the gas-oil ratio does not change.
- **Stock-tank gravity (API):** as gravities of retrograde gas condensate reservoir and reach above 60° API. Also during wet gas reservoir life, stock-tank gravity of condensate liquid remains constant.
- **Color:** water-white.

**Table 2** shows data for three different examples of wet gas reservoirs.

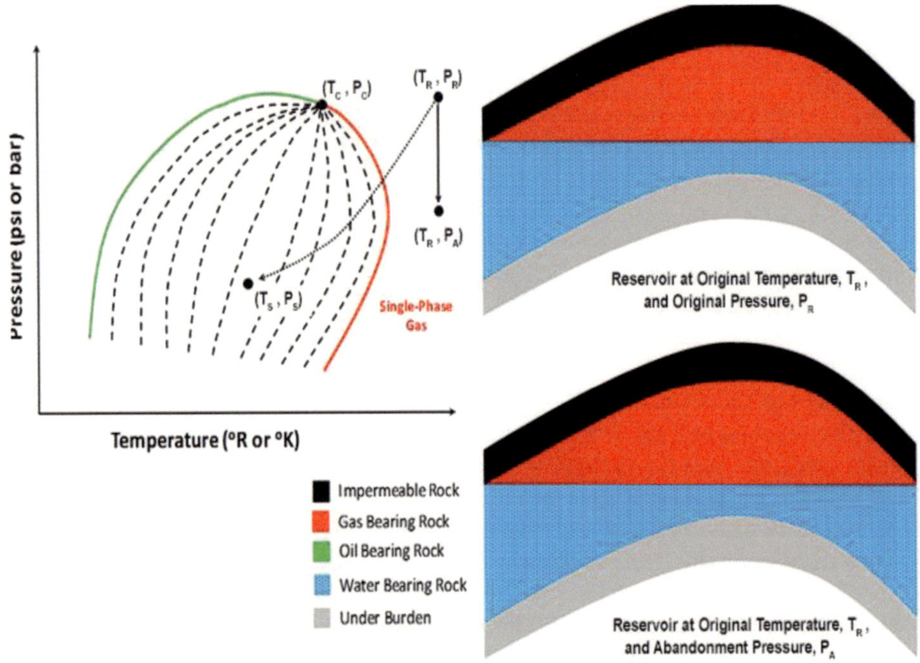


**Figure 3.**
*Wet gas reservoir phase diagram.*

| **Reservoir information** | **Sample 1** | **Sample 2** | **Sample 3** |
|---|---|---|---|
| Reservoir pressure, psi | 1248.52 | 6040 | 9118.3 |
| Reservoir temperature, °F | 120 | 286 | 221 |
| $C_7^+$, Mole % | 0.3494 | 0.3806 | 0.8714 |
| Average mole weight | 20.4650 | 21.14 | 19.40 |
| GOR, STB/SCF | 238,346.92 | 360,000.00 | 67,142.857 |
| API | 54.800 | 58.55 | 47.88 |
| **Compositional analysis of reservoir fluid to C36+** | | | |
| **Component** | **Mole %** | **Mole %** | **Mole %** |
| Hydrogen | 0.0000 | 0.0000 | 0.0000 |
| Hydrogen sulfide | 0.0000 | 0.0000 | 0.0000 |
| Carbon dioxide | 0.1450 | 4.8233 | 0.3852 |
| Nitrogen | 0.0232 | 0.0639 | 0.0018 |
| Methane | 84.6734 | 78.7860 | 90.9414 |
| Ethane | 5.5664 | 10.6768 | 3.9810 |
| Propane | 5.8349 | 3.5138 | 2.0867 |
| i-Butane | 0.8776 | 0.3639 | 0.4953 |
| n-Butane | 1.5846 | 0.6540 | 0.5146 |
| Neo-Pentane | 0.0000 | 0.0000 | 0.0000 |
| i-Pentane | 0.5351 | 0.2036 | 0.2605 |
| n-Pentane | 0.0855 | 0.1961 | 0.0880 |
| Hexanes | 0.1991 | 0.1784 | 0.2267 |
| M-C-Pentane | 0.0446 | 0.0311 | 0.0390 |
| Benzene | 0.0199 | 0.0670 | 0.0369 |
| Cyclohexane | 0.0606 | 0.0609 | 0.0714 |
| Heptanes | 0.0801 | 0.0478 | 0.1015 |
| M-C-Hexane | 0.0501 | 0.0545 | 0.0434 |
| Toluene | 0.0384 | 0.1240 | 0.0425 |
| Octanes | 0.0479 | 0.0337 | 0.0993 |
| E-Benzene | 0.0100 | 0.0048 | 0.0040 |
| M/P-Xylene | 0.0248 | 0.0239 | 0.0281 |
| O-Xylene | 0.0145 | 0.0028 | 0.0107 |
| Nonanes | 0.0234 | 0.0160 | 0.0600 |
| 1,2,4-TMB | 0.0040 | 0.0003 | 0.0056 |
| Decanes | 0.0185 | 0.0153 | 0.0662 |
| Undecanes | 0.0128 | 0.0121 | 0.0642 |
| Dodecanes | 0.0052 | 0.0096 | 0.0599 |
| Tridecanes | 0.0030 | 0.0069 | 0.0475 |
| Tetradecanes | 0.0026 | 0.0059 | 0.0452 |
| Pentadecanes | 0.0022 | 0.0048 | 0.0388 |
| Hexadecanes | 0.0016 | 0.0041 | 0.0348 |
| Heptadecanes | 0.0014 | 0.0035 | 0.0241 |
| Octadecanes | 0.0013 | 0.0026 | 0.0217 |
| Nonadecanes | 0.0014 | 0.0020 | 0.0169 |
| Eicosanes | 0.0012 | 0.0016 | 0.0137 |

| Component | Mole % | Mole % | Mole % |
|---|---|---|---|
| Heneicosanes | 0.0014 | 0.0011 | 0.0120 |
| Docosanes | 0.0013 | 0.0009 | 0.0088 |
| Tricosanes | 0.0008 | 0.0008 | 0.0042 |
| Tetracosanes | 0.0009 | 0.0006 | 0.0051 |
| Pentacosanes | 0.0004 | 0.0005 | 0.0026 |
| Hexacosanes | 0.0004 | 0.0004 | 0.0022 |
| Heptacosanes | 0.0002 | 0.0003 | 0.0016 |
| Octacosanes | 0.0001 | 0.0002 | 0.0014 |
| Nonacosanes | 0.0001 | 0.0001 | 0.0009 |
| Triacontanes | 0.0001 | 0.0001 | 0.0008 |
| Hentriacontanes | 0.0001 | 0.0001 | 0.0007 |
| Dotriacontanes | 0.0001 | 0.0001 | 0.0005 |
| Tritriacontanes | 0.0000 | 0.0000 | 0.0003 |
| Tetratriacontanes | 0.0001 | 0.0000 | 0.0002 |
| Pentatriacontanes | 0.0001 | 0.0000 | 0.0000 |
| Hexatriacontanes plus | 0.0000 | 0.0000 | 0.0020 |
| Total | 100.00 | 100.00 | 100.00 |

**Table 2.**
*Examples of wet gas reservoirs.*

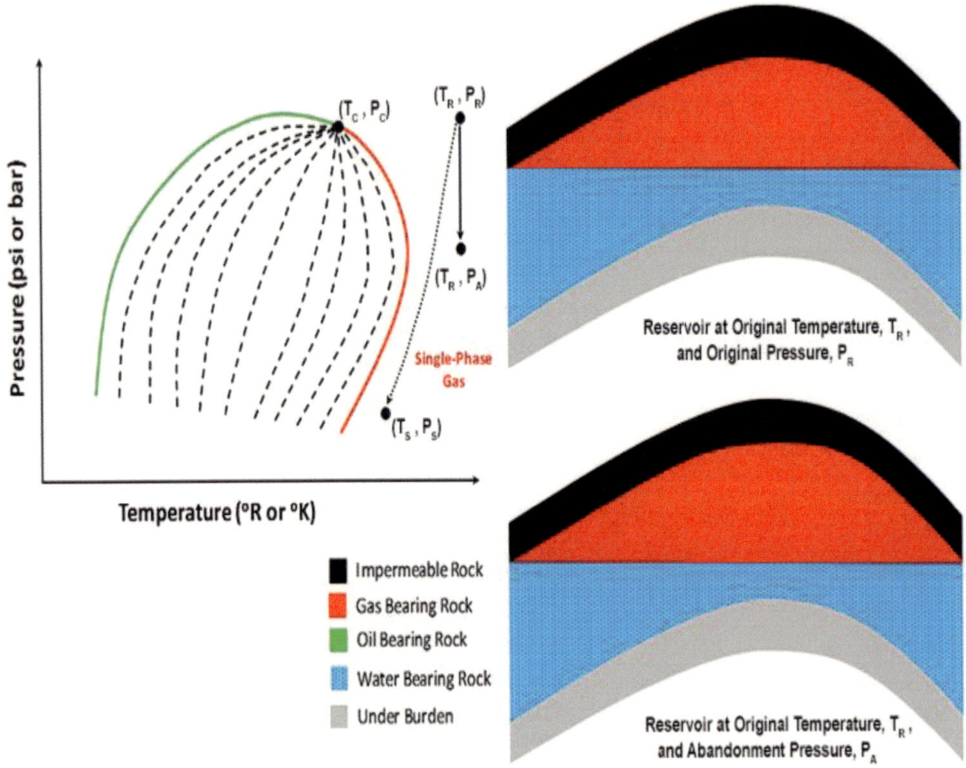


**Figure 4.**
*Dry gas reservoir phase diagram.*

| **Reservoir information** | **Sample 1** | **Sample 2** | **Sample 3** |
|---|---|---|---|
| Reservoir pressure, psi | 6754 | 7853 | 6545 |
| Reservoir temperature, °F | 234 | 330 | 301 |
| $C_7^+$, Mole % | 0.0370 | 0.0175 | 0.0101 |
| Average mole weight | 16.6817 | 16.455 | 16.57 |
| GOR, STB/SCF | 320,000.00 | 270,000.00 | 232,100.00 |
| **Compositional analysis of reservoir fluid to C36+** | | | |
| **Component** | **Mole %** | **Mole %** | **Mole %** |
| Hydrogen | 0.0000 | 0.0000 | 0.000 |
| Hydrogen sulfide | 0.0000 | 0.0000 | 0.000 |
| Carbon dioxide | 0.4370 | 0.4112 | 0.081 |
| Nitrogen | 0.1450 | 0.0054 | 0.012 |
| Methane | 97.5760 | 98.2135 | 97.812 |
| Ethane | 0.9540 | 0.9099 | 1.102 |
| Propane | 0.3520 | 0.2989 | 0.615 |
| i-Butane | 0.2190 | 0.0306 | 0.152 |
| n-Butane | 0.1540 | 0.0547 | 0.113 |
| Neo-Pentane | 0.0000 | 0.0000 | 0.000 |
| i-Pentane | 0.0560 | 0.0164 | 0.054 |
| n-Pentane | 0.0240 | 0.0158 | 0.015 |
| Hexanes | 0.0350 | 0.0134 | 0.022 |
| M-C-Pentane | 0.0060 | 0.0024 | 0.003 |
| Benzene | 0.0020 | 0.0055 | 0.003 |
| Cyclohexane | 0.0030 | 0.0046 | 0.004 |
| Heptanes | 0.0140 | 0.0025 | 0.003 |
| M-C-Hexane | 0.0070 | 0.0031 | 0.004 |
| Toluene | 0.0050 | 0.0094 | 0.002 |
| Octanes | 0.0030 | 0.0011 | 0.001 |
| E-Benzene | 0.0010 | 0.0003 | 0.000 |
| M/P-Xylene | 0.0010 | 0.0011 | 0.000 |
| O-Xylene | 0.0010 | 0.0000 | 0.000 |
| Nonanes | 0.0020 | 0.0000 | 0.000 |
| 1,2,4-TMB | 0.0010 | 0.0000 | 0.000 |
| Decanes | 0.0020 | 0.0000 | 0.000 |
| Undecanes | 0.0000 | 0.0000 | 0.000 |
| Dodecanes | 0.0000 | 0.0000 | 0.000 |
| Tridecanes | 0.0000 | 0.0000 | 0.000 |
| Tetradecanes | 0.0000 | 0.0000 | 0.081 |
| Pentadecanes | 0.0000 | 0.0000 | 0.012 |
| Hexadecanes | 0.0000 | 0.0000 | 97.812 |
| Heptadecanes | 0.0000 | 0.0000 | 1.102 |
| **Component** | **Mole %** | **Mole %** | **Mole %** |

| Component | Mole % | Mole % | Mole % |
|---|---|---|---|
| Nonadecanes | 0.0000 | 0.0000 | 0.152 |
| Eicosanes | 0.0000 | 0.0000 | 0.113 |
| Heneicosanes | 0.0000 | 0.0000 | 0.000 |
| Docosanes | 0.0000 | 0.0000 | 0.054 |
| Tricosanes | 0.0000 | 0.0000 | 0.015 |
| Tetracosanes | 0.0000 | 0.0000 | 0.022 |
| Pentacosanes | 0.0000 | 0.0000 | 0.003 |
| Hexacosanes | 0.0000 | 0.0000 | 0.003 |
| Heptacosanes | 0.0000 | 0.0000 | 0.004 |
| Octacosanes | 0.0000 | 0.0000 | 0.003 |
| Nonacosanes | 0.0000 | 0.0000 | 0.004 |
| Triacontanes | 0.0000 | 0.0000 | 0.002 |
| Hentriacontanes | 0.0000 | 0.0000 | 0.001 |
| Dotriacontanes | 0.0000 | 0.0000 | 0.000 |
| Tritriacontanes | 0.0000 | 0.0000 | 0.000 |
| Tetratriacontanes | 0.0000 | 0.0000 | 0.000 |
| Pentatriacontanes | 0.0000 | 0.0000 | 0.000 |
| Hexatriacontanes plus | 0.0000 | 0.0000 | 0.000 |
| Total | 100.00 | 100.00 | 100.00 |

**Table 3.**
*Examples of wet gas reservoirs.*

## 2.3 Dry gases reservoirs

This type is a gas phase in the reservoir and in the surface condition, where surface separator conditions located outside the phase envelope as given in **Figure 4** [17]. This diagram also shows that no liquid is formed at stock-tank condition (temperature and pressure) as a result of no attraction between molecules. This type also simply called a gas reservoir. Dry gas is mainly methane component with some intermediates components. The expression of "dry gas" refers to does not have heavier molecules to form condensate liquid at the surface condition. In this case, gas-oil ratios are reached more than 100,000 SCF/STB [18]. **Table 3** shows data for three different examples of wet dry reservoirs.

# 3. Conclusion

This chapter converses the hydrocarbon fluids phase behavior. The physical properties of these three natural gas reservoir fluids, with emphasis on retrograde gas condensate gas, dry gas, and wet gas. The behavior of reservoir is determined by phase diagram shape and critical point position. All examples show the details of each fluid type by reservoir information and compositional analysis of reservoir fluid.

## Author details

Eman Mohamed Mansour*, Mohamed El Aily and Saad Eldin Mohamed Desouky
PVT Lab, Production Department, Egyptian Petroleum Research Institute,
Cairo, Egypt

*Address all correspondence to: emanmansour84@yahoo.com

## References

[1] Mansour E, Desouky S, Batanoni M, Mahmoud M, Farag A, El-Dars F. Modification proposed for SRK equation of state. Oil and Gas Journal. 2012;**110**(6):78-91

[2] Guo B, Ghalambor A. Natural Gas Engineering Handbook. Gulf Professional Publishing: Elsevier; 2014

[3] Mansour E, Farag A, El-Dars F, Desouky S, Batanoni M, Mahmoud M. Predicting PVT properties of Egyptian crude oils by a modified Soave–Redlich–Kowng equation of state. Egyptian Journal of Petroleum. 2013;**22**(1):137-148

[4] Bradley HB. Petroleum Engineering Handbook. Society of Petroleum Engineers; 1987

[5] Mansour E, Al-Sabagh A, Desouky S, Zawawy F, Ramzi M. Experimental approach of minimum miscibility pressure for $CO_2$ miscible flooding: Application to Egyptian oil fields. International Journal of New Technology and Research. 2016;**2**(5):105-112

[6] Ahmed T. Equations of State and PVT Analysis. Gulf Professional Publishing: Elsevier; 2013

[7] Mansour E, Al-Sabagh A, Desouky S, Zawawy F, Ramzi M. A new estimating method of minimum miscibility pressure as a key parameter in designing $CO_2$ gas injection process. Egyptian Journal of Petroleum. 2017;**27**:801-810

[8] Mansour E, Al-Sabagh A, Desouky S, Zawawy F, Ramzi M. A laboratory investigation of carbon dioxide-enhanced oil recovery by focusing on $CO_2$-oil physical properties. Egyptian Journal of Petroleum. 2019;**28**:21-26

[9] El Aily M, Khalil M, Desouky S, Batanoni M, Mahmoud M. Experimental studies on constant mass–volume depletion of gas-condensate systems. Egyptian Journal of Petroleum. 2013;**22**(1):129-136

[10] El Aily M, Desouky S, Khalil M, Badawi A, Abdelraheem O. Effect of dry gases injection into Egyptian gas/condensate reservoirs on condensate recovery. International Journal of New Technology and Research. Nextgen Research Publication. **2**:119-121

[11] Masoumi S, Helalizadeh A, Bahrami M. The performance of various injecting gases into fractured retrograde gas reservoirs for revaporization of liquid drop-out. Petroleum Science and Technology. 2011;**29**(24):2536-2544

[12] McCain WD Jr. Properties of Petroleum Fluids. PennWell Publishing: PennWell Corporation; 2017

[13] Afidick D, Kaczorowski N, Bette S. Production performance of a retrograde gas reservoir: A case study of the Arun field. In: SPE Asia Pacific Oil and Gas Conference; Society of Petroleum Engineers; 1994

[14] Ahmed T. Reservoir Engineering Handbook. Gulf Professional Publishing; 2018

[15] James AT. Correlation of reservoired gases using the carbon isotopic compositions of wet gas components (1). AAPG Bulletin. 1990;**74**(9):1441-1458

[16] Boyer C, Kieschnick J, Suarez-Rivera R, Lewis RE, Waters G. Producing gas from its source. Oilfield Review. 2006;**18**(3):36-49

[17] McCain W Jr. Reservoir-fluid property correlations-state of the art (includes associated papers 23583 and 23594). SPE Reservoir Engineering. 1991;**6**(02):266-272

[18] Lee WJ, Wattenbarger RA. Gas Reservoir Engineering. 1996

Chapter 6

# Damage Formation: Equations of water block in oil and water wells

*Mohammad Karimi, Mohammad Reza Adelzadeh, Mojtaba Mosleh Tehrani, Maryam Mohammadipour, Ruhangiz Mohammadian and Abbas Helalizade*

## Abstract

Water block or invasion of water into the pores of reservoir forms during the operations of water-based drilling, injection, many perforations, completion fluids, and some other particular processes in the reservoir (such as fingering and conning). Subsequently, the alteration in the shape or composition of the fine particles such as clay (water-wet solids), as a result of the stress on it, in the flow path of the second phase can lead to the permeability decline of reservoir. Consequently, the solvents such as surfactants (as demulsifiers) to lower the surface tension as a phenomenon associated with intermolecular forces (known as capillary action) during flowback are consumed to avoid the emulsions and sludge mostly in the near-wellbore zone or undertreatment and under-injection radius of the reservoir. However, in addition to surging or swabbing the wells to lower the surface tension, using solvents as the wettability changing agent along with base fluid is a common method in the water block elimination from the wellbore, especially in the low permeability porous media or the reservoirs latter its average pressure declined below bubble point. For more profitability, after using solvents in various reservoir characterizations, the trend of their behavior variations in the different lithologies is required to decide on the removed damage percentage. The investigations on this subject involve many experimental studies and have not been presented any mathematical formulas for the damage of water block in the water, oil, and gas reservoirs. These formulas determine selection criteria for the applied materials and increase variable performance. An integrated set of procedures and guidelines for one or more phases in a porous media is necessary to carry out the step-by-step approach at wellhead. Erroneous decisions and difficult situations can also be addressed in the injection wells or saltwater disposal wells, in which water block is a formation damage type. Misconceptions and difficult situations resulting from these injuries can increase water saturation in borehole and affect the fluid transmissibility power in reaching far and near distances of the wellbore, which results in injection rate loss at the wellhead. Accordingly, for the equations of water block here, a set of variables, of a particular domain, for defining relationships between rock- and fluid-based parameters are required. For these equations, at first, the structural classifications of fracture and grain in the layers ($d_1$, $d_2$, and $d_3$) are defined. Afterward, the equations of overburden pressure ($P_{ob}$) for a definite sectional area surrounding the wellbore for any lithology (in the three categories relative to porosity) are obtained by these structural classifications and other characteristics of rock and fluid. Naturally, prior to equations of overburden pressure in



IntechOpen

a definite layer or a definite sectional area around the wellbore, the overburden pressure of a point in a layer in the first four equations is expressed. In the second, the estimated overburden pressure equations are applied in driving the equations of removed water block ($B_k$). The equations of removed water block, themselves, are divided into two groups of equations, i.e., equations of oil wells and equations of saltwater disposal wells, and each group of equations is again classified based on the wettability of reservoir rock (oil-wet or water-wet) in the two ranges of porosity. In the third, after describing these equations (i.e., equations of $B_k$), the other new variable included in the equations of removed water block, that is, the acid expanding ability ($I_k$) for a definite oil layer around the wellbore, is presented, which is extracted from (1) the full characteristics of reservoir (including experimental and empirical equations of overburden pressure), (2) the history of producing well, (3) core flooding displacement experiments at laboratory, and (4) the acidic and alkaline solvent properties. Finally, the rate of forming water block (q) is calculated using the value calculated for the removed water block, and, additionally, the trend of using solvents is determined for different rocks using these sets of equations. The acceptance criteria are the nature of rock and fluid in the reservoir circumstances. Equations as a quick and cost-efficient method are also introduced, providing computational methods to determine how much and how the blocked fluid in the reservoir layers is removed from the definite strata around the wellbore after injection operation of acids and solvents, with various degrees of acidity, to the types of lithology during acidizing operations. Moreover, these equations can calculate the removed water block ($B_k$) after injecting solvents to the different acidic properties in the acidizing, for two categories of porosity which cover all lithologies. The equations also ascertain in the current reservoir conditions how much solvent for a type of lithology is to be mixed with other base fluids.

**Keywords:** defined water and oil layer overburden pressure, overburden pressure of a definite layer point, removed water block equations, injection wells or saltwater disposal wells (SWDW), oil wells, rock and fluid characteristics, chemical solvents

## 1. Literature review

For avoiding productivity loss in wells, the compatibility of lithology with types of acids, their use percentage, and additive solvents mixed with base fluids are preferred to be handled before acidizing. In the downhole operations, use of solvents such as alcohols and surfactants (as wetting factors for lowering surface tension of the acid and subsequently for better penetration in the matrix of rock) should be carried out in accord with the previously-estimated quantities for any lithology so that the wettbility changes provide the stable conditions for the engaging phases and control losing of oil–based phases toward the formation. The work on water block in the previous literature is in the form of experimental investigations (Holditch, 1999; [1–3]), and an integrated method is necessary for forecasting the outcome of the interactions related to fluid and rock after injecting acids and any fluid mixed with the wetting agents at which our purpose is to dissolve the water blocked in the oil wells. For this, however, finding the exact rock characteristics and the data on injected and in situ fluid behavior in the reservoir is imperative to accurately derive the equations of water block. In other words, the equations of water block are also introduced so as to present computational methods to find out how much and how the blocked fluid is removed from the definite strata with the specific lithology, during injecting solvents.

## 2. Quantitative structural characteristic classification table in reservoir

Three physical quantities, i.e., intergranular space (IGS), inter-fracture space (IFS), and fracture width (FW), are given in **Table 1**. To find the magnitude of these three variables for mathematical expressions, we consider the effect of fine-grained particles' migration severity on porosity and the affect of carbonate cement or clay in the layer on the damages such as water block, phase trapping and any other obstacle caused by rock and fluid [4]. Generally, too much attention is given to information of porosity and permeability in the qualitative and quantitative situations, including cementing, color, compaction pressure, consolidating and unconsolidating property, particle size in lithology, density, and distances of fracture and grain experimentally and empirically to both the oil layers (oil-wet and water-wet types) and the saltwater disposal wells (mostly water-wet).

### 2.1 Example

For a reservoir layer containing sand associated with dolomite, the data of intergranular space ($d_1$), inter-fracture space ($d_2$), and fracture width ($d_3$) using **Table 1** is obtained (see a thin section of the whole plug in **Figure 1** which has the $\varphi$ of 16%).

**Solution:** The variables $d_1$ and $d_2$ are obtained on averaging the values given in their related ranges to each group of rocks in **Table 1** (**Figure 2** illustrates how to figure out $d_1$, $d_2$, and $d_3$ in a sample of reservoir layer). Note: As you know, although

| ID | Lithology | IGS, m | ID | Lithology | IGS, m | Symbols |
|---|---|---|---|---|---|---|
| 1 | CP | $5 \times 10^{-2}$ to $10^{-2}$ | 5 | L/D | $6.5 \times 10^{-5}$ to $7.5 \times 10^{-6}$ | S: sand<br>D: dolomite<br>L: limestone<br>SH: shale |
| 2 | Fine S | $10^{-2}$ to $2.5 \times 10^{-3}$ | 6 | S with SH | $7.5 \times 10^{-6}$ to $10^{-7}$ | |
| 3 | S with L/D | $2.5 \times 10^{-3}$ to $5.5 \times 10^{-4}$ | 7 | SH/clay | $<10^{-7}$ | |
| 4 | S with L/D/SH | $5.5 \times 10^{-4}$ to $6.5 \times 10^{-5}$ | | Inter-granular space: (IGS) or $d_1$ | | |
| **ID** | **Lithology** | **IFS, m** | **ID** | **Lithology** | **IFS, m** | |
| 1 | | $2.5 \times 10^{-2}$ to $10^{-2}$ | 5 | L/D | $10^{-5}$ to $10^{-6}$ | -Inter-granular space: (IGS) or $d_1$<br>-Inter-fracture space: (IFS) or $d_2$.<br>-Fracture width: (FW) or $d_3$, |
| 2 | Fine S | $10^{-2}$ to $10^{-3}$ | 6 | S with SH | $10^{-6}$ to $10^{-7}$ | |
| 3 | S with L/D | $10^{-3}$ to $5.5 \times 10^{-4}$ | 7 | SH/clay | $<10^{-7}$ | |
| 4 | S with L/D/SH | $5.5 \times 10^{-4}$ to $10^{-5}$ | | Inter-fracture space: (IFS) or $d_2$ | | |
| **ID** | **Lithology** | **FW, m$^2$** | **ID** | **Lithology** | **FW, m** | |
| 1 | CP | $<10^{-7}$ | 5 | S with L/D | $10^{-4}$ to $10^{-5}$ | Coarse particles: CP |
| 2 | CP | $10^{-6}$ to $10^{-7}$ | 6 | S with SH | $10^{-3}$ to $10^{-4}$ | |
| 3 | Fine S | $5.5 \times 10^{-5}$ to $10^{-6}$ | 7 | L/D | $10^{-2}$ to $10^{-3}$ | |
| 4 | S with L/D/SH | $5.5 \times 10^{-4}$ to $10^{-5}$ | 8 | Sh/clay | $>10^{-2}$ | |
| | | | | Fracture width: (FW) or $d_3$ | | |

**$d_1$ and $d_2$ are the average per range in each ID. For $d_3$ we have $d_3$ = 100% (0.1 M), $5 < \varphi \le 15.5$; $d_3$ = 30% (0.1 M), $20.5 < \varphi \le 25$; $d_3$ = 50% (0.1 M), $15.5 < \varphi \le 20.5$, M = ($d_{max}$-$d_{mim}$/$d_{max}$).*

**Table 1.**
*Quantitative structural characteristic classification table in reservoirs.*

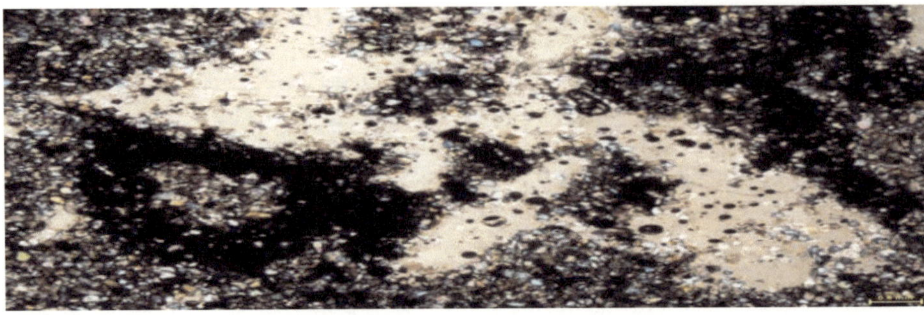

**Figure 1.**
*A fairly porous typical oil layer drilled from 3572 m and synthetic sandstone with less percentage of fine crystals of anhydrite, dolomite cement, and sub-round/fossil patches. This rock is associated with small ooids in the middle and intraclasts on all sides which have been poorly sorted. The porosity type is fracture and enormous vugs ($\varphi$ = 16%) containing the oil.*

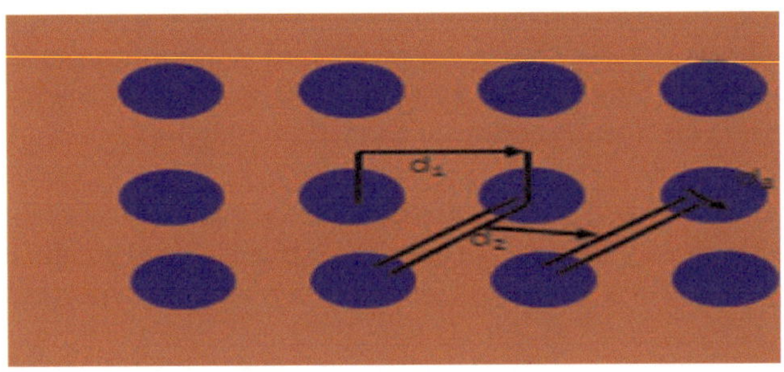

**Figure 2.**
*Situation of grains and fractures to each other in matrix media.*

the dolomite and limestone are not the same, all their characteristics, to some extent excluding chemical properties, are mostly similar. Therefore, to calculate $d_1$, $d_2$, and $d_3$ for either dolomite or limestone, we apply a quantity defined for dolomite/limestone in the classification table (**Table 1**):

$$d_1 = \frac{2.5 \times 10^{-3} + 5.5 \times 10^{-4}}{2} = 0.00153 \; d_2 = \frac{10^{-3} + 5.5 \times 0^{-4}}{2} = 0.0008$$

$$M = \frac{10^{-4}\text{-}10^{-5}}{10^{-4}} = 0.9 \quad \text{For } 15.5 < \varphi \leq 20.5, \text{ we have } d_3 = 50\%(0.1\,M) = 0.045.$$

## 3. Formation damage and overburden pressure

This subsection summarily described how overburden pressures, as formulas below, affect the formation damage. The variable can influence the other physical parameters (e.g., porosity and permeability) and underground interactions. Due to its great variability, carbonate rocks are the most onerous to construe and analyze, and their pressure can also range in various levels and change the fluid distribution in the pores. Hence, the overburden pressure influences physical parameters, especially, as the pressure in depths alters the fluid movements and the tectonic displacements. The experimental overburden pressure results on these physical parameters demonstrate a decrease in porosity and permeability while rising overburden pressure in reservoirs [5, 6]. All these variations are observed in the "equation of overburden pressure" [4] in the next sections. In addition, overbalance pressure can affluence a couple of processes in the drilling operations [7].

Overburden pressure can lead to the forming of water block that subsequently will alter the porous media and its fluid, for example, in the reservoirs with dual medium, permeability reduction resulted from overburden pressure of reservoir can prepare the media condition for forming water block [8, 9]. The forces, such as overburden pressure, which changes many physical characteristics in the section of formation adjacent to the wellbore, can relate to the subsurface processes that displace particles while fluids flow through propose media. These forces exerted by fluid and rock on the small drilled or damaged point or on the major underground dimensions could naturally alter the layer pressure and other physical characteristics [10, 11]. Therefore, in small drilled or damaged point or major underground dimensions, the classification of overburden pressure based on porosity is used in the "equations of removed water block."

### 3.1 Equation of overburden pressure

#### *3.1.1 Equation of overburden pressure in a definite point of a layer*

Some methods to calculate overburden pressure are according to Eqs. (1)–(4). Equation (1) is presented by Hubbert and Rubey (1959) for overburden pressure at a depth z that is a function of parameters z, $P_0$, and g:

$$P(z) = P_0 + g\int_0^z \rho(z)dz \tag{1}$$

where ρ (z) and z are, respectively, the density of the overlying rock and depth and g is the acceleration due to gravity. $P_0$ is the same datum pressure. Another useful equation for calculating overburden gradient of varying lithology and pore fluid density (this formula can calculate the pressure in every depth) is derived by Matthews and Kelly [12]:

$$\sigma_{ovg}=0.433\left[(1-\phi)\rho_{ma}+\left(\rho_f\phi\right)\right] \tag{2}$$

where $\sigma_{ovg}$ is the overburden gradient, psi/ft., $\varphi$ is the porosity expressed as a fraction, and $\rho_{ma}$ is the formation fluid density, gr/cc.

Another method presented by Karimi et al. [4, 13, 14] contains six equations summarized in Eqs. (3) and (4), respectively, for oil layers and water layers. In the equations of (3) and (4), the information on petrophysics and geology and the quantitative-structural characteristics classification table the reservoir conditions (**Table 1**), $P_{ob}$ formulated for a point of drilled layer in various porosity ranges. Since reservoir layers have mostly the heterogeneity characterizations, geologists and drillers need to control timely and repeatedly the type of cuttings, drilling mud, and reservoir pressure; the equations can help to effectively accomplish the operations in wellhead and bottom hole through the calculations in which the overburden pressure is important:

$$P_{ob1} = \frac{1}{A}\left|\frac{C_1\rho_r g h A_w\sqrt{\left(\frac{\mu_o+\mu_w}{\mu_w}\right)}\sqrt{\frac{t_1}{t}}}{{d_1}^2}\right| + \frac{1}{A}\left|\frac{C_2(W_W+W_o)}{}\frac{1}{{d_3}^2}\right| = P_{ob1,1}+P_{ob1,2} \tag{3}$$

$$C_1 = 2\times10^{-5}; C_2 = 0.02; d_3 = 100\%(0.1\,M)\ 5<\varphi\leq15.5$$

$$C_1 = 1.75 \times 10^{-5}; C_2 = 0.04; d_3 = 50\%(0.1\,M)\ 15.5 < \varphi \le 20.5$$

$$C_1 = 1.5 \times 10^{-5}; C_2 = 0.06; d_3 = 30\%(0.1\,M)\ 20.5 < \varphi \le 25$$

$$P_{ob1} = \frac{1}{A}\left|\frac{C_1\rho_r g h A_w\sqrt{\left(\frac{\mu_w@60°F - \mu_w@T°F}{\mu_w}\right)}\sqrt{\frac{t_1}{t}}}{d_1^{\,2}}\right| + \frac{1}{A}\left|C_2(W_W)\frac{1}{d_3^{\,2}}\right| = P_{ob1,1} + P_{ob1,2} \quad (4)$$

$$C_1 = 2 \times 10^{-5}; C_2 = 0.25; d_3 = 100\%(0.1\,M)\ 5 < \varphi \le 15.5$$

$$C_1 = 1.75 \times 10^{-5}; C_2 = 0.3; d_3 = 50\%(0.1\,M)\ 15.5 < \varphi \le 20.5$$

$$C_1 = 1.5 \times 10^{-5}; C_2 = 0.35; d_3 = 30\%(0.1\,M)\ 20.5 < \varphi \le 25$$

### 3.1.2. Equations of overburden pressure in a definite layer

The equations of overburden pressure of "a definite and delimited sectional area" that contains three equations, summarized in Eq. (5), are included in the equations of the removed water block ($B_k$) in the oil wells (in the next sections); hence, the removed water block is a function of the overburden pressure of a definite and delimited sectional area. More details on the equation of overburden pressure have been given in references. The variations of overburden pressure for three groups of rocks are given in Eq. (5):

$$P_{ob} = \frac{1}{A}\left|\frac{C_1\sqrt{W_d}\sqrt{\rho_r g h}\sqrt{\left(\frac{\mu_o + \mu_w}{\mu_w}\right)}\sqrt{\frac{t_1 + t}{(t_1 - t)}}}{d_1\sqrt{\frac{d_1^{\,2}}{d_2}}}\right| + \frac{1}{A}\left|C_2(W_W + W_o)\frac{d_2}{d_1 d_3^{\,2}}\right| + \frac{1}{A}\left|\frac{C_3 W_d}{d_3^{\,2}}\right| \quad (5)$$

where

$$C_1 = 0.2;\ C_2 = 1.10;\ C_3 = 4.6 \times 10_{+2}\ d_3 = 100\%(0.1M);\ 5 < \phi \le 15.5$$

$$C_1 = 0.17; C_2 = 1.15; C_3 = 4.4 \times 10^{+2}\ d_3 = 50\%(0.1M);\ 15.5 < \phi \le 20.5$$

$$C_1 = 0.15;\ C_2 = 1.20;\ C_3 = 4.3 \times 10^{+2}\ d_3 = 30\%(0.1M);\ 20.5 < \phi \le 25 M = \left(\frac{d_{\max} - d_{mim}}{d_{\max}}\right)$$

$$\text{If } T < 176°F \text{ then } \mu_o@T° = \mu_0\text{-}22pH_0@60°F \times \frac{(176°F - T°)}{60°F}[(\rho_o@60°\text{-}\rho_o@T°)/\rho_o@60°] \text{ and } \mu_0 = \mu_o@60°F, \quad (6)$$

$$\text{If } T \ge 176°F \text{ then } \mu_o@T° = \mu_0\text{-}17pH_0@60°F \times \frac{(T° - 176°F)}{60°F}[(\rho_o@60°\text{-}\rho_o@T°)/\rho_o@60°] \text{ and } \mu_0 = \mu_o@176°F, \quad (7)$$

where $P_{ob}$ is the formula of overburden pressure (bar) and the constant A is a unit conversion factor from $k_{gf}$ to bar and equals to 10197.162. $\varphi$ indicates porosity (%). $\rho_r$ is the rock density (kg/m$^3$), and the $\rho_o$ and $\rho_w$ are the oil and water densities (kg/m$^3$), respectively. $\rho_r = W_d/(V_b\text{-}V_p)$ at which $V_b$ is the bulk volume (m$^3$), $V_p$ is the empty space volume ($m^3$), and $W_d$ is the dry weight for sectional area without any fluid ($k_{gf}$). g is the acceleration of gravity($kg/m^3$), h is the layer depth from earth surface (m), and t is the geological age of favorite layer on million years (my), at which $t_1$ denotes the lower layer age. $d_1$ and $d_2$ denote, respectively, intergranular space and inter-fracture space (it refers to matrix media or distance between fractures) (m), while $d_3$ denotes fracture width (m). $d_{max}$ and $d_{mim}$ in Eq. (5) determine the maximum and minimum fracture width in each lithology which is calculated from **Table 1**. $C_1$, $C_2$, and $C_3$ are defined as the constants in which $C_1$ and $C_3$ are associated with $W_d$ (dry layer weight) and their values, and also their effect on $P_{ob}$ is maximum in the oil layers with low porosity, but $C_2$ is associated with the fluid weight ($W_w$ + $W_o$), i.e., oil and water, and its volume, and also its effect on $P_{ob}$ becomes minimum in the low-porosity oil layers. In general, the values of constants change for different layers in the determined porosities, depending on the composition of layer. $\mu_o$ is obtained by Eqs. (6) and (7), which indicates the viscosity of oil (kg/m-sec) in an oil layer at reservoir temperature (°F), and pH is oil acidity which usually does not change in the reservoir media.

### *3.1.3. Exercise*

In this exercise you will be familiarized with the method of obtaining overburden pressure in "a definite point of an oil layer with length of 80 m in depth 3593 m." This layer is composed of sand with a little dolomite and anhydrite, and the geological period is pre-Miocene (25 my ago), which is located on a layer with geological age of Eocene period (50 my ago). The geological and petrophysical characteristics of layer ($\varphi$ = 18.86%) are given below. (A) Use Eqs. (2) and (3) to solve the problem. (B) If, instead of a definite point, the goal is to treat a definite layer or delimited sectional area (i.e., whole sectional area with l = 80 m and A = $1.13 \times 10^{+3}$ $m^2$), then use Eq. (5) to solve the problem. (C) Assume that there is a definite layer or delimited sectional area ($\varphi$ = 11% and with lithology of sand associated with less percentage of shale in depth of 3605 m) with geological age of Eocene period (50 my ago) with the same size exactly beneath the layer with geological period of pre-Miocene (25 my ago). (D) The lowest sectional area with geological age of 65 my with the same size is exactly beneath Eocene period. (Note on C: use **Table 1** (sand with shale) to calculate d ($d_1$, $d_2$, and $d_3$) and then $P_{ob}$, assuming that the other data is identical with other sectional areas). The sectional areas are depicted with A, B, C, and D in **Figure 3**.

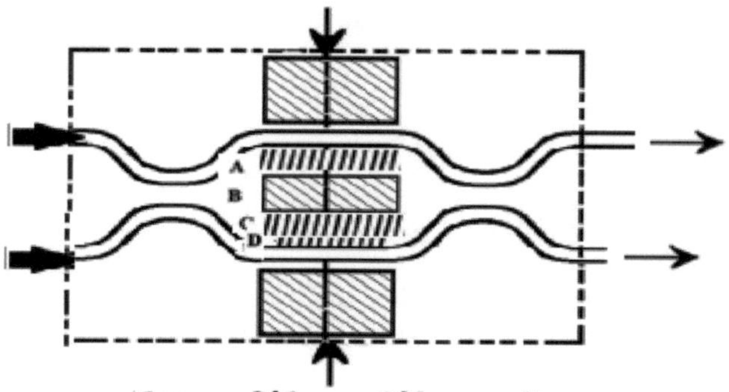


**Figure 3.**
*A schematic of four reservoir layers with various geological ages.*

Characteristics of the sub-layer 2 of formation A: Cylindrical area section n of layer A ($m^2$) = $1.13 \times 10^{+3}$; $A_{we}$ = 0.05 $m^2$; $W_d$ ($k_{gf}$) = 2,038,984; h (m) = 3593; $\rho_w$ at 60°F = $1.145 \times 10^{+3}$; $\rho_w$ at 191°F = $1.135 \times 10^{+3}$; $\rho_o$ at 60°F = 851.9; $\rho_o$ at 191°F = 821.9; $\rho_r$ ($kg/m^3$) = $2.82 \times 10^{+3}$; K (md) = 23; t (my) = 25; $t_1$ (my) = 50; L (m) = 80; $\mu_o$ at 60° F = 11.3; $\mu_o$ at 176°F = 3.3; $\mu_o$ at 191°F = 2.43; $pH_o$ at 60°F = 6.8; $\mu_w$ at176°F = 0.72; $\mu_w$ at 60°F = 1.73; $\mu_w$ at 191°F = 0.65 cp; $d_1 = 7.75 \times 10^4$; $d_2 = 5.5 \times 10^4$; $V_w$ ($m^3$) = 3.37; $V_o$ ($m^3$) = 13.49; $V_d$ ($m^3$) = 73.78; T (°F) = 191. Due to the patches of anhydrite in containing of rock, $d_1$ may be close to $d_2$ (density unit is $kg/m^3$).

## 4. Water block and formation

In this section the characteristics of layer such as rock transmissibility, effect of pathways on the fluid conductivity and particles, and the other physical variables in underground conditions are studied. The previously carried out studies discern the water block subject matter from the other damages, like the phase trapping index (APTi). In the studies investigators researched on the role of oil-based fluids, surface tension, injection of dry gas, drawdown pressure, and outputs of displacement and evaporation processes on the trend of blocking fluid around the wellbore within the fracture and matrix. For example, see [15–21]. But many studies are carried out on the solvents and water block, and their results determined the role of solvents and other fluids injected on injection type and fingering and coning water in water, oil, and gas reservoirs around the wellbore as well as their destructive effect on reservoir through water block [20–22]. However, the water block as a major issue in water injection wells can damage permeability more in sandy layers, and clay minerals become sensitive to the salinity degree and pH of water injected [23]. In low-permeability reservoirs, the containing of reservoir can displace the clay minerals [24]. The changes in wettability can improve the water block depending on the reservoir conditions [25]; in this case some experimental works are designed and carried out to the wettability alterations on the liquids [26]. Moreover, to discern the interactions of rock and fluid, other important researches are carried out for the dense and loose reservoirs by [1, 2]. The modeling results on the fractured and unfractured systems by Parekh [27] and their production times by Lake [28] highlighted the considerable water block in fractures due to the less pressure difference between the capillary and drawdown states. Consequently, in the case of the difference between static and flowing bottom-hole pressures, the fracture effect has an important role in decision-making on the overburden pressure as well as the type and the percentage use of solvent used in any lithology [13, 14].

## 5. Selection criteria

In driving the "Equations of Water block" and "Equation of Overburden Pressure" it is necessary to determine whether the heterogeneity alterations widely studied and monitored in certain periods through reports during our equations have any positive effect on product performance in defining main thresholds for each variable. However, the criteria should be set based on (1) the rock and fluid nature at the reservoir conditions, (2) the intended use of the wellhead product history, and (3) most importantly the previously delimited area around the wellbore that injection in a great deal time slowly will improve it [4, 29–36].

### 5.1. General methodology for novel equations

the discussions below composed of (1) the use of **Table 1**, (2) the use of equation of overburden pressure, and (3) the introduction of equations of water block for the oil-wet and water-wet layers of oil reservoirs for removing the damage.

## 6. Equations of removed water block ($B_k$) in oil reservoirs

### 6.1. Equations of $B_k$ in the oil-wet oil reservoirs with dimensions defined

The equations of removed water block have been obtained through the experimental data and empirical information to answer to many fundamental questions on anisotropic and isotropic reservoirs that their composition includes multi mineral with various pores geometry and varied relative proportions. The equations proposed solutions to contrast the various oil layers and presented the alternatives to treat their damage while confronting to the damage in the porous media. The equations of oil-wet oil reservoirs for two porosity ranges are expressed in Eqs. (8) and (9):

$$B_k = \sqrt{k}q_i\sqrt{-\frac{I_k A_w \Delta P}{\frac{\rho_f V}{t}(q_{\max}-q_{\min})}}\sqrt{(\mathrm{pH_f}/\mathrm{pH_{ac2\%}})\text{-}1}\sqrt{(1-\frac{D_s}{H_{Mg}})^2}\quad I_k = 6.5\times 10^2(\mathrm{T^o})\ (h)\left(1\text{-}\frac{\mu_f}{\mu_{ac2\%}}\right)^4\left(\frac{P_p}{P_{ob}{}^{C_1}}\right)^{-2}$$

$$C_1 = 0.10 - C_{mu};\ 0.05 < C_{\mathrm{mu}} \le 0.10\ ;5 < \phi \le 15.5 \tag{8}$$

$$(B_k)_{o.w} = \sqrt{k}q_p\sqrt{-\frac{I_k A_w \Delta P}{\frac{\rho_f V}{t}(q_{\max}-q_{\min})}}\sqrt{\left(\frac{\mathrm{pH_f}}{\mathrm{pH_{ac4\%}}}\right)\text{-}1}\sqrt{\left(1-\frac{D_S}{H_{Mg}}\right)^2}\quad I_k = 7.1\times 10^2(\mathrm{T^o})(h)\left(1\text{-}\frac{\mu_f}{\mu_{ac4\%}}\right)^4\left(\frac{P_p}{P_{ob}{}^{C_1}}\right)^{-2}$$

$$\mathrm{C_1} = 0.10\text{-}\mathrm{C_{mu}};\ 0.05 < \mathrm{C_{mu}} \le 0.10\ ;15.5 < \varphi \le 25 \tag{9}$$

### 6.2. Equations of $B_k$ in the water-wet oil reservoirs with dimensions defined

The variables mentioned above for oil-wet reservoirs are applied here for the water-wet oil reservoir equations. These equations for two ranges of the porosity are expressed in Eqs. (10) and (11) as follows:

$$(B_k)_{w.w} = \sqrt{k}q_p\sqrt{-\frac{I_k A_w \Delta P}{\frac{\rho_f V}{t}(q_{\max}-q_{\min})}}\sqrt{\left(\frac{\mathrm{pH_f}}{\mathrm{pH_{ac2\%}}}\right)\text{-}1}\sqrt{\left(1-\frac{D_S}{H_{Mg}}\right)^2}$$

where

$$I_k = 5.4\times 10^2(\mathrm{T^o})(h)(1\text{-}\mu_f/\mu_{ac2\%})^4\left(\frac{P_p}{P_{ob}{}^{C_1}}\right)^{-2}\quad C_1 = 0.10 - \mathrm{C_{mu}};\ 0.05 < \mathrm{C_{mu}} \le 0.10; 5 < \varphi \le 15.5 \tag{10}$$

$$(B_k)_{w.w} = \sqrt{k}q_p\sqrt{-\frac{I_k A_w \Delta P}{\frac{\rho_f V}{t}(q_{\max}-q_{\min})}}\sqrt{\left(\frac{\mathrm{pH_f}}{\mathrm{pH_{ac4\%}}}\right)\text{-}1}\sqrt{\left(1-\frac{D_S}{H_{Mg}}\right)^2}$$

where

$$I_k = 5.9 \times 10^2(\mathrm{T}^{\circ})(h)\left(1\text{-}\frac{\mu_f}{\mu_{ac4\%}}\right)^4\left(\frac{P_p}{P_{ob}{}^{C_1}}\right)^{-2} \mathrm{C}_1 = 0.10 - \mathrm{C}_{mu}; 0.05 < \mathrm{C}_{mu} \le 0.10; 15.5 < \varphi \le 25 \quad (11)$$

where removed water block or $B_k$ is the power of chemical in the damage removal and proves the relationship between the expanding of chemicals and the damage made. $B_k$ is proportional to the rate of blocking the fluid ($q_B$) and reversely to the square root of the acid expanding ability ($\sqrt{I_k}$) in which $q_B$ is calculated from $q = \frac{Bk}{\sqrt{I_k}}$. In the $B_k$ the volume of water block that per minute endured the temperature T° in depth of h after injecting acid and solvents with diverse acidity properties, is estimated, and its unit is $\frac{m^3}{\min\sqrt{T^{\circ}m}}$. $I_k$ is the acid expending ability, in which its unit becomes mK or T°m, where 1 K = 1000 mK. One K denotes the expending of an acid sample of 28% injected to a well with the previously predicted pressure and rate of injection and production. And the well is located in a layer L of centimeter long (a length of zone around the wellbore which the front of the fluid injected covers which zone and causes the damage) with a cylindrical cross section at depth h from the earth surface which endures an overburden pressure of its upper layer column. Depending on the favorite layer lithology, the acid injected with base fluids is converted to acids 2–4% (ac2% to ac4% in Eqs. (10) and (11)) at reservoir circumstances. In these conditions, the viscosity of base fluid mixed with solvent and the expended acid viscosity are assessed for reservoir fluid displacement behavior. Practically, 1 K measures the capacity of the gradual expending of an acid in which its variable is indicated with $I_k$ and is applied where the sectional areas with large horizontal scales of the reservoir layers programmed to treat, stimulate, fracture, complete, and/or drill. In the expending acid, the media with a more ability can push the previously blocked fluid with rate of $q_B$, in which the underground chemical expending is measured and assessed with $I_k$. $C_{mu}$ determines the percent solvent used associated with base fluid that in oil reservoirs is mainly gasoil, and the constant of $C_1$ is calculated by which. The percent of $C_m$ is commonly being used at wellhead and in relationship of $C_1$ is subtracted from the maximum allowable amount delimited by factory's product. At equations, this maximum amount becomes 0.10 for mutual solvents. Accordingly, the superscript $C_1$ varies in value for the solvent types that subsequently will change the variable of $I_k$. $P_{ob}$, overburden pressure (bar), is obtained from Eqs. (1)–(3). K is the permeability (md), and SI has unit of K which is equal to about $0.98692 \times 10^{-12}$ or $10^{-12}$ $m^2$. $q_p$ is the last oil rate ($m^3$/min) in production well before injecting fluids. $P_p$ is the last oil pressure (bar) in production well before injecting fluids. ΔP is the pressure loss of acid 28% and retarder acid with gasoil mixed in solvent (bar), and $A_w$ is considered the sectional area of wellbore ($m^2$). V equals the entire volume of fluids injected into the well ($m^3$), excluding the volume of fluid mixed with solvent. t is the injection time of the entire volume of fluids injected into the well (min). $\rho_f$ is the density of base fluid (kg/$m^3$) in which mutual solvent is mixed with it at wellhead. $\mu_f$ is the μ of base fluid mixed with solvent that is equal to the average of viscosity of the base fluid (in here is gasoil) and viscosity of the solvent (kg/m-s) under reservoir conditions. $\mu_{ac4\%}$ is the viscosity of acid 27% and retarder acid (kg/m-s) that have endured the conditions of reservoir. pH of the base fluid mixed with solvent is the average of the acidity percent of the base fluid and mutual solvent under reservoir conditions. $pH_{ac4\%}$ is the acidity of acid 27% and retarder acid, which have endured the reservoir conditions in a sandy layer and have been converted to acid 4%, as in most of the limestone layers this amount is 2–3% (this value in its related equations averagely based with ac2%). $D_s$ and $(H)_{Mg}$ are, respectively, salinity degree (ppm) and Mg hardness of formation water (ppm), and according to the procedure of oil industry to obtain the Mg hardness, we apply the $Ca^{+2}$ to 0.4 ratio. The sensitivity and alteration range of variables in equations depends on the reservoir nature.

### 6.3. Expending ability or $I_k$

$I_k$ denotes acid expending ability based on mK, which generally demonstrates the viscosity for acids expended to 2 and 4% under reservoir conditions for the two groups of the limestone/dolomite layers and the sandstone layers, as the other groups change between these percentages. For example, a chemical in a limestone media with high temperature and permeability can better be expended than a sandy media usually with low porosity; instead, in the same sandy media, the ($B_k$) volume is more. The $I_k$ represents not only the property of the acid, but also it describes the various rocks that could alter in form and structure by natural agents. See **Figure 4** which is only to indicate the trend of alterations in the variables of the correlation of $I_k$ for the different layers. Also, **Figure 4** illustrates the acid expending ability versus overburden pressure in the reservoir layers which have directly been analyzed from the field and lab information. The exact $I_k$ is obtained from the correlation of $I_k$ for types of the reservoir lithology; thus, we cannot extrapolate the curves to obtain a special variable so that the intercept is read as an exact value. In general, as the $\varphi$ increases, $C_m$ or the same percentage solvent used also increases. Subsequently, $C_1 = 1 - C_m$ in the $P_{ob}^{C1}$ declines and the $I_k$ also declines. Under these conditions, $q_B$ or the rate of blocking is high and $B_k$ (or $q_B \sqrt{I_k}$) also increases. As was discussed, the $B_k$, power of damage removal by chemical, is directly proportional to the product of the square root of expending in the rate of blocking; in other words, $q_B$ is the expending square root of the removed water block.

The $q_B$ obtains through the ratio of the $B_k$ obtained from the equation to $\sqrt{I_k}$. Most often, as the h increases, then $\varphi$ also decreases. Subsequently, the $C_m$ or the same percentage solvent decreases, and the superscript $C_1$ in both $P_{ob}^{C1}$ and $I_k$

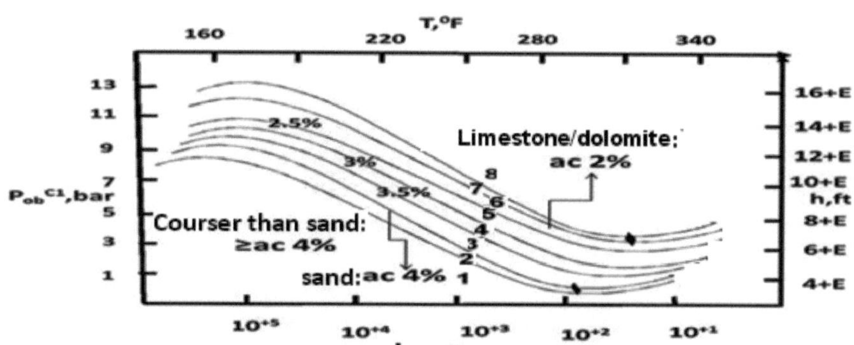


| ID | Lithology | ID | Lithology |
|---|---|---|---|
| 1 | Coarse particles | 5 | Sand with limestone /dolomite |
| 2 | Coarse sands | 6 | Shale sands |
| 3 | Fine sands | 7 | Limestone/dolomite |
| 4 | Sand with limestone /dolomite/shale | 8 | Shale/clay |
| ac% in 2 as 3 and ac% in 7 as 8 (although in 8 a little less, including $P_{ob}^{C1}$ in $I_k$ is ignored ); To estimate T° and h using $P_{ob}^{C1}$ and $I_k$ obtained of formulas, don't use of this curves. | | | |

**Figure 4.**
*Acid expending ability versus overburden pressure in reservoir layers [33, 36, 37].*

increases. Under this condition in which the acid expending ($I_k$) is more, up to 2%, $q_B$ or rate of blocking in such a low-porosity media with relatively usual permeability is low, and the amount of ($B_k$) is less. If the $I_k$ is estimated for a horizontal layer to a length larger than 100 m, then we can estimate the $I_k$ together with other layers as a multiple of 100 m (e.g., for the pure sandy layer in length of 152 m, the entire $I_k$ is multiple of 1.52. If we assume that 52 m of this layer is limestone and 100 m is sand, then the entire $I_k$ is the sum of 0.52 $I_k$ and 1 $I_k$ in which the data are substituted in the related correlations to $I_k$). For additional detailed information on the index of $I_k$, the reader is referred to the section of units.

### 6.3.1. Exercise

An oil well with pressure of 600 psi is produced from the oil-wet sandy layer (containing a little dolomite) with porosity 18.86%, temperature 191°F, and length 80 m, located in depth 3593 m. The acid 28% injected to increase the production, after 204.83 min using sampling from backflow at wellhead, determined the acid in the reservoir conditions expanded to an acid 4% and its viscosity is 0.55. In the next step, the percent solvent used with base fluid to reduce surface tension during operations is 4.5% whole base fluid (nearly less than 5 bbl), and allowable percent delimited for solvent by factory is in the range of $0.05 < C_{mu} \leq 0.09$. In the laboratory at 191°F, the fluid viscosity ($\mu_f$) is measured and calculated from the average viscosities of solvent and water (nearly 1.21 cp). Estimate $I_k$ using Eq. (8) for this reservoir if $P_{ob}$ obtained through Eqs. (5)–(7) for this definite sectional area is 880,224,837 bar.

#### 6.3.1.1. Challenges in equations

The equations presented in this capture have not been interpreted for the gas-bearing layers. In the gas reservoirs, the pore spaces are quite smaller relative to spaces saturated to the water and oil in the oil and water reservoirs. As writing these equations for gas-bearing layers, the following should be noted: (1) the diversity and size of porous in gas-bearing layers is less than that in oil-bearing layers or as such in water-bearing type. For this, formulating of related equations for overburden pressure does not need that they be categorized into the three groups of porosity as conducted above for the oil layers or the water layers. Instead, devising an equation for any porosity less or equal to 5 is sufficient. (2) Since the gas compressibility varies considerable as compared to the density in the liquids, thus only the property of gas compressibility relative to its density caused the gas molecules to occupy the less space that consequently will increase the effects of adhesion and cohesion. As a result, the measure of “high viscosity” for the liquids will be modified for the gas viscosity when the compressibility of gas as a significant variable is included in its related “single equation” for the viscosity. And (3) a combination of porosity and compressibility in the modified equation of gas viscosity proves the equation of overburden pressure for gas layers in a range of porosity (1–5%) that then can correct the equations of ($B_k$) to the gas reservoirs. Accordingly, in the equation of viscosity that has been corrected through compressibility for a gas layer’s overburden pressure, the variation range of porosity in Eq. (5) in the set equations on overburden pressure ranges from 1 to 5% instead of 5 to 10%, and the term d or $d_1$, $d_2$, and $d_3$, from **Table 1**, can distinguish the type of lithology for any gas reservoir in which the equation is modified.

Now, if a researcher intends to demonstrate the equations of water block ($B_k$) for the gas well, it needs to derive only one equation according to the corrections on compressibility at the abovementioned discussions. In a single equation, afterward, the porosity would be modified and applied to calculate the overburden pressure of a defined layer around the wellbore. However, only two equations, ultimately, can predict the ($B_k$) in the gas wells to the two wettability types of water-wet and oil-wet (although the gaseous layers become mostly water-wet). As a final note, I must say that if there is no any water or oil in a gas reservoir, then we can assume the value of "viscosity" in the equation of overburden pressure is zero.

## 7. Example

The goal in this example is to obtain the rate of removed water block ($B_k$) formed in a definite layer or sectional area (cylindrical type) around an oil well with oil-wet wettability. The lithology of delimited layer is sandy containing dolomite, which is located in the depth 3593 m at 191°F (φ,% = 18.83). To treat the damage, at first the gasoil mixed with solvent (mutual type) is consumed in the acidizing. Using the given date below, calculate $P_{ob}$ and $B_k$. The salinity degree, Ca ion of formation, and viscosity of fluids are obtained from **Tables 2** and **3**.

Solution problem:

Geology data of layer (L = 80 m): {A ($m^2$) = $1.13 \times 10^{+3}$; $W_d$ ($k_{gf}$) = 203,8984; L (m) = 80; h (m) = 3593}, {$\rho_w$ at 60°F = $1.145 \times 10^{+3}$; $\rho_w$ at 191°F = $1.135 \times 10^{+3}$; $\rho_o$ at 60°F = 0.8519; $\rho_o$ at 191°F = 0.8219; $\rho_r$ (kg/$m^3$) = $2.82 \times 10^{+3}$}, {φ (%) = 18.86; k (md) = 23; t (my) = 25; $t_1$ (my) = 50}, {$\mu_o$ at 60°F = 11.3; $\mu_o$ at 176°F = 3.3; $\mu_o$ at 191°F = 2.43; $\mu_w$ at 176°F = 0.72; $\mu_w$ at 60°F = 1.73; $\mu_w$ at 191°F = 0.65 cp; $pH_o$ at 60°F = 6.8}, {$d_1 = 7.7 \times 10^{+4}$; $d_2 = 5.5 \times 10^{+4}$; $d_3$ = ?}, {$V_w$ ($m^3$) = 3.37; $V_o$ ($m^3$) = 13.49; $V_d$ ($m^3$) = 73.78; T (°F) = 191}. Because the patches of anhydrite in containing of rock, the $d_1$ equals to $d_2$.

Calculation $P_{ob}$: In **Table 1** in ID of 4, we have $d_3$ = 50% (0.1 M) = $(d_{max}–d_{mim})/d_{max}$; $d_3{}^2 = [0.05(5.5 \times 10^{-5}–10^{-5}) /(5.5 \times 10^{-5})]^2 = 0.002$ m. $\mu_w$ at T = 191°F using its related equations equals to 2.43 cp. If we substitute the data in equation of overburden, then $P_{ob}$ in a layer with length of 80 m in depth of 3593 m equals to 834,678,036 + 409 + 455,466,392 = 8 80,224,837 bar.

Fluid and well data: $\rho_{water}$ at 191°F = 8620 kg/$m^3$; $\rho_{ac4\%}$ at 60°F = 1180; $\rho_{solvent}$ at 191°F = 910; $\mu_{ac4\%}$ at 176°F = 0.57; $\mu_{ac4\%}$ at 191°F = 0.55; $\mu_{water}$ at 60°F = 1.43; $\mu_{solvent}$ at 191°F = 0.99; $\mu_{water + solvent}$ at 191°F = 1.21; $pH_{ac4\%}$ at 191°F = 0.67; $pH_{water}$ at 191°F = 5.7; $pH_{solvent}$ at 60°F = 4.99; $pH_{solvent}$ at 191°F = 7.83; $pH_{water + solvent}$ at 191°F = 6.77; $P_{water} = P_1$ = 1700 psi = 115.6 bar; $P_{water + solvent}$ $P_2$ = 950 psi = 64.5 bar; $P_p$ = 600 psi = 41 bar; $q_{mim} = q_{water + solvent}$ = 4.5 bbl/min = (0.72 $m^3$); $q_{mam} = q_{water}$ = 9.5 bbl/min = (1.51 $m^3$); $q_p = q_i$ = 1.4 bbl/min; {$r_w$ = 0.42 ft. = 0.128 m; $A_{we}$ = 0.05 $m^2$; $V_{we}$ = 1187.25 bbl}, {$V_{water}$ = 775 bbl; $V_{solvent}$ = 5.5 bbl; $V_{water + solvent}$ = 100 bbl; $V_p$ = 105,090 bbl; V = $V_{inj} = V_{water}$ = 775 bbl = 123.22 $m^3$};

| **Hardness, ppm** | | **Salinity** | | | **Ions, ppm** | | | |
|---|---|---|---|---|---|---|---|---|
| Total | Ca | 220,000 | Cl | Mg | Ca | Fe | Co3 | So4 |
| 54,000 | 45,000 | | 150,875 | 2187 | 18,000 | 74 | 854 | 425 |
| pH of Water | | | pH of low-viscosity solvent (LVS) | | | pH of high-viscosity solvent (HVS) | | |
| 5.4 | | | 4.99 | | | 9.52 | | |

**For the filtration of water, the filter 0.45 μm was used. The pH of water is before boiling.*

**Table 2.**
*Formation water ionic specifications.*

| Fluid type | T (°F) | μ (cp) | Type and constant of tube used |
|---|---|---|---|
| Oil | 60 | 11.93 | s.2, 0.006 |
| | 176 | 3.3 | |
| | 191 | 2.43 | |
| | 220 | 2.35 | |
| | 225 | 2.09 | |
| Gasoil | 60 | 5.06 | Cannon-Fenske |
| | 176 | 1.59 | |
| | 191 | 1.43 | |
| | 220 | 1.24 | |
| | 225 | 1.20 | |
| Water | 60 | 1.73 | s.50, 0.004247 |
| | 176 | 0.75 | |
| | 191 | 0.62 | |
| | 220 | 0.52 | |
| | 225 | 0.51 | |
| LVS | 60 | 4.47 | s.1.c, 0.03102 |
| | 176 | 1.39 | |
| | 191 | 0.99 | |
| | 220 | 0.81 | |
| | 225 | 0.76 | |
| HVS | 60 | 7.09 | s.1, 0.01345 |
| | 176 | 1.73 | |
| | 191 | 1.04 | |
| | 220 | 0.76 | |
| | 225 | 0.71 | |

*$\rho_w$ = 1.145, $\rho_o$ = 0.8519, $\rho_{gasoil}$ = $\rho_{go}$ = 0.8620, $\rho_{LVS}$ = 0.91, $\rho_{HVS}$ = 0.8910, oil salt = 14; ρ is g/cm$^3$; LVS, low-viscosity solvent; HVS, high-viscosity solvent.*

**Table 3.**
*Typical fluid viscosity specifications used/compared in various temperatures.*

$D_s$ = 220,000 ppm; $H_{Mg}$ = 9000 ppm; $\rho_{water}$ at 191°F = 8620; $C_{mu}$ = 5.5%; $t_{inj}$ = 204.83 min.

Calculation $(B_k)_{O.W}$: C = 0.10-$C_{mu}$ = 0.1–0.055 = 0.045; $P_{ob}{}^{C}$ = 2.52 bar; $(P_p/ P_{ob}{}^{C})$ = 16.27; $(1-\mu_f/\mu_{ac4\%})^2 = 1-(1.2/0.55)^2$ = 1.44; $[(pH_f/pH_{ac4\%})-1]^{0.5}$ = 3.02. Using $\rho_{go}$, $V_{inj}$ in loose oil-wet rock and Δp = −65.6 bar, we have $[(A_w\Delta P)/((\rho_f V/t) \times (q_{max}-q_{mim}))]^{0.5} = 0.03(1/m^2)^{0.5} = 0.03(1/mmDarcy)^{0.5}$, and using equation of $I_k$ at 191°F, now $I_k$ equals 3705 mK. If we substitute the data in equation of $B_k$ for loose oil-wet rocks, then we have $(B_k)_{O.W} = 52.53(m^3/min) \times [(T°m)^{-0.5}]$.

In the wells produced from a layer with silt compositions, the high salinity and calcium ion concentration have a considerable role in the increase of calcium deposits around the wellbore and formation of water block, as these percentages are higher in saltwater disposal wells. With this method of calculation, we can measure the damage for the other wells with the various characteristics of formation that have been previously producing a constant flowing bottom-hole pressure (BHP) and now have confronted to the damaging or unloading conditions and or the production rate loss.

## 8. Equations of water block in saltwater disposal wells: a chemical injection process for removing damage in saltwater disposal wells

The acids diluted with water, due to the membrane of water on the rock and the penetration of water into the pores, prevent the immediate contact of the acid against rock as well as high diluting cannot also strength against dense minerals. Therefore, the solvents (as demulsifiers) especially mutual solvents as a mediocre fluid are injected associated with the various percentages of acids to control the reactions formed by the water block on the rock at a smaller scale. In practice, the application of these investigation is based on the used solvents in acidizing to help (1) provide an integrated solution of removing the water block in a definite sectional area around the SWDW's wellbore using the properties of the rock and the injected and in situ fluids (e.g., solution type, rock composition, solids size, pressure, concentration, etc.), (2) control the parameters of rock and fluid at the wellhead and reservoir as well as predict how and where these parameters are able to be controlled, (3) provide the computational methods with most measures in which these methods determine how much of water block is removed in the reservoir layers with various lithologies in the course of injecting solvents with the various acidity properties in the acidizing operations, (4) provide the methods that enable the designer to match and manipulate the occurrences inside the reservoir rock before starting the injection operations and as such enable them for recognition of its treatment, and (5) facilitate the software applications at the time of access to the state-of-the-art facilities and various producing chemicals to the disciplined and methodical approach that for this aim: (a) these equations associated with the equations related to oil layers in other references without any onerous technique are easy to code up, (b) this paper and similar it to the oil wells in other references that would help in writing the software and/or sub-equations of these equations in the calculations of above-mentioned underground processes (stimulation, fracturing, recovery, capturing, etc.).

Before anything, for this aim and better understanding of the effect of the chemical on damage, a process of forming and removing the water block in the water-wet oil-bearing rocks is shown in **Figure 5a–c**, and then the theory of water block in the formation is presented. This schematic indicated the trend of forming and removing damage before and after using water, interfacial tension (IFT), and mutual solvent (MUS).

As discussed above in the introduction, in the methodology, first, the structural layer characteristics (IGS, IFS, and FW) are obtained from the experiments on the oil-and water-wet layers and the wellhead information mentioned (**Table 1** in previous pages); second, the equation of overburden pressure as a function of the physical parameters resulted from previous information indicates its own role in

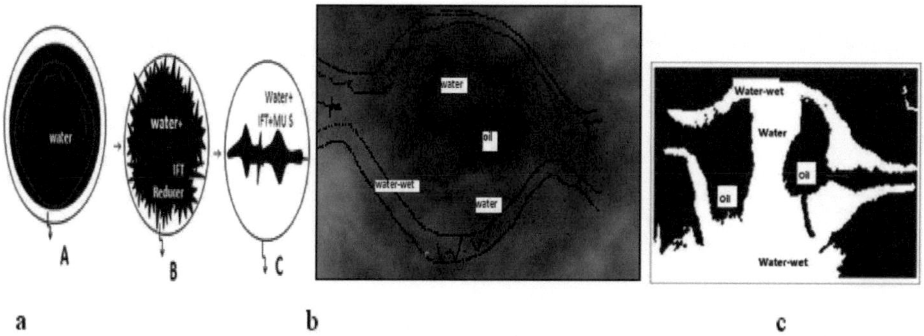


**Figure 5.**
*(a) Schematic of the mutual solvents treatment process under reservoir conditions: (A), primary situation of water-wet reservoir contains water; (B) injection of the IFT reducer, dash link area added to oil-wet part; (C) injection of mutual solvent through conductivity and miscibility caused the mixture of chemical and water to flow outward the matrix, and then the replacing fluid should be injected with low speed to prevent the damage of matrix. (b) An image of forming water block. (c) A schematic of the water block after developing in the path of the residual oil flow [13, 14].*

revealing parameters effecting on the water block (Eqs. (8)–(10)); and finally, the novel equations of water block are introduced for the water-wet layers of injection wells.

## 8.1 Equations of overburden pressure in the water layers

As mentioned above, the main subject matter is related to a definite stratum in which its dimensions are specified, and we aim to break/diminish the blocked water. Thus, first, the overburden pressure determined in a definite sectional area [13, 14, 37] through six equations (summarized here in Eqs. (10)–(12)), is applied to include all structural and non-structural characteristics of rock mass, lithology at various times and dynamic fluid distribution in the removed water block equations.

$$P_{ob} = \frac{1}{A}\left|\frac{C_1\sqrt{w_d}\sqrt{\rho_r g h}\sqrt{\frac{\mu_w@60^\circ - \mu_w@T^\circ}{\mu_w@T^\circ}}\sqrt{\frac{t+t_1}{(t_1-t)}}}{d_1{}^2\sqrt{\frac{1}{d_2}}}\right| + \frac{1}{A}\left|C_2 \times W_w \times \frac{1}{d_1 \times d_3{}^2} \times d_2\right| + \frac{1}{A}\left|\frac{C_3 \times W_d}{d_3{}^2}\right| \tag{12}$$

$C_1 = 1.11; C_2 = 1.1; C_3 = 9 \times 10_{+2}\ d_3 = 100\%(0.1M) \quad 5 < \varphi \le 15.5$

$C_1 = 0.99;\ C_2 = 1.15;\ C_3 = 8.7 \times 10^{+2}\ d_3 = 50\%(0.1M) \quad 15.5 < \varphi \le 20.5$

$C_1 = 0.90;\ C_2 = 1.20;\ C_3 = 8.6 \times 10^{+2}\ d_3 = 30\%(0.1M) \quad 20.5 < \phi \le 25\ M = \frac{d_{\max} - d_{mim}}{d_{\max}}$

Water velocity in the reservoir is determined using.

$$\text{If } T < 176°F,\ \mu_w@T° = \mu_0\text{-}0.12pH_w@60°F \times \frac{(176°F - T°)}{60°F}[(\rho_w@60°\text{-}\rho_w@T°)/\rho_w@60°]$$

where

$$\mu_0 = \mu_w@60°F \tag{13}$$

$$\text{If } T \ge 176°F,\ \mu_w@T° = \mu_0\text{-}0.09pH_w@60°F \times \frac{(T° - 176°F)}{60°F}[(\rho_w@60°\text{-}\rho_w@T°)/\rho_w@60°]$$

where

$$\mu_0 = \mu_w@176°F \tag{14}$$

### *8.1.1. Units in $P_{ob}$*

$P_{ob}$ is overburden pressure (bar) in which the constant of A is to convert the unit of $k_{gf}$ to the bar and equals to 10197.162. h is the depth of layer from earth surface (m), φ is porosity (%), and g is the acceleration of gravity (kg/m$^3$). $\rho_r$ is rock density, kg/m$^3$, and $\rho_w$ is water density, kg/m$^3$. $\rho_r = W_d/(V_b\text{-}V_p)$ at which $V_b$ = bulk volume (m), $V_p$ is pore volume (m$^3$) and $W_d$ is dry weight ($k_{gf}$). t is geological age of favorite layer on the million years (my) at which $t_1$ is the lower layer age. $\mu_w$ is viscosity of water contact in water layer under reservoir temperature (kg/m-s), and the pH is water acidity. T is the reservoir or experiment condition temperature (°F). $C_1$, $C_2$, and $C_3$ are dimensionless constants at which the $C_1$ and $C_3$ relate the $W_d$ (dry layer weight) in which their value and also their effect on $P_{ob}$ in the layer with low porosity are maximum, whereas $C_2$ relates the fluid weight ($W_w + W_o$) in which its value and also its effect on $P_{ob}$ in low porosity layers are minimum. Therefore, the constant values would change with different sets of rocks in the determined porosities. $d_1$ and $d_2$ are, respectively, intergranular space and inter-fracture space (matrix media or distance between fractures) on meters. $d_3$ is fracture width on meters. $d_{max}$ and $d_{mim}$ are maximum and minimum fracture width in each lithology which is obtained from **Table 1.**

## 8.2 Equations of removed water block in saltwater disposal wells ($B_k$ in SWDW)

The damage of water block is one of the formation damages caused by the increase of water saturation in the near or very far distances from wellbore and can occur either in the oil wells or saltwater disposal/depleted oil wells [13, 14, 33, 36]. To remove the damage, the solvents are usually used (especially mutual solvents) associated to other base fluids such as water and gasoil in the treatment processes of the water and oil wells [38–43]. In this investigation the equations are presented to estimate the removed water block in the salt water disposal wells according to Eqs. (15) and (16):

$$B_k = \sqrt{k}q_i\sqrt{-\frac{I_k A_w \Delta P}{\frac{\rho_f V}{t}(q_{max}-q_{min})}}\sqrt{(\mathrm{pH_f}/\mathrm{pH_{ac2\%}})\text{-}1}\sqrt{(1-\frac{D_s}{H_{Mg}})^2}$$

$$I_k = 1.7\times 10(\mathrm{T}^\circ)(h)\left(1\text{-}\frac{\mu_f}{\mu_{ac2\%}}\right)^4\left(\frac{P_i}{P_{ob}{}^{C_1}}\right)^{-2}$$

$$C_1 = 0.10 - C_{mu};\ 0.03 < C_{mu} \le 0.10;\ 5 < \varphi \le 15.5 \quad (15)$$

$$B_k = \sqrt{k}q_i\sqrt{-\frac{I_k A_w \Delta P}{\frac{\rho_f V}{t}(q_{max}-q_{min})}}\sqrt{(\mathrm{pH_f}/\mathrm{pH_{ac4\%}})\text{-}1}\sqrt{\left(1-\frac{D_s}{H_{Mg}}\right)^2}$$

$$I_k = 1.8\times 10(\mathrm{T}^\circ)(h)\left(1\text{-}\frac{\mu_f}{\mu_{ac4\%}}\right)^4\left(\frac{P_i}{P_{ob}{}^{C_1}}\right)^{-2}$$

$$C_1 = 0.10 - C_{mu};\ 0.03 < C_{mu} \le 0.10;\ 15.5 < \varphi \le 25 \quad (16)$$

The unmentioned units of variables here for Eqs. (15) and (16) are the same as in Eqs. (8)–(11). The most percent of solvents removing damage, for example mutual solvents ($C_{mu}$), is commonly used in gas wells and at least in water wells. If chemical mixed to base fluid is alcohol, surfactant, and/or any other reaction controller, then the changes of $C_{mu}$ depend upon well conditions and usage range in the chemical catalog. $P_{ob}$ (on the bar), in Eqs. (15) and (16), is the overburden pressure that is obtained from Eqs. (12)–(14). $q_i$, $m^3$/min, is the last water rate in the SWDW before injecting fluids, and $P_i$, bar, is the last injection pressure before injecting fluids. ΔP, bar, is the pressure difference of acid 28% and retarder acid with water mixed in the solvent. V, $m^3$, is the entire volume of fluids injected to the well, excluding the volume of fluid mixed with solvent. t, min, is the injection time of entire volume of fluids injected to the well. $\mu_f$ is the density of base fluid mixed with solvent. $\mu_f$, kg/m-s, is the average of viscosity of base fluid (in here is water) and solvent mixed with it under reservoir conditions obtained at laboratory. $\mu_{ac4\%}$, kg/m-s, is the viscosity of acid 27% and retarder acid that have endured conditions of reservoir. $pH_f$ is the average of acidity for base fluid mixed with solvent under reservoir conditions obtained at laboratory.

### 8.3. Layer thickness in equations

Thickness determines the pressure in the lowest layer in which depth h is treated and endures the weight of the reservoir column. The real value of thickness is its impact on the lowest reservoir layer. Strictly speaking, this is a complementary application of h that is used to calculate the overburden pressure and/or any other quantity such as formation damage (e.g., it is the type of water block), mud optimum pressure, and petrophysical and geological parameters of the reservoir in the processes of stimulation, acidizing, micro-fracturing, and recovery in domain of the definition of reservoir layers. It has been observed in equations that a depth (h) at the large size for overburden pressure could not become the same depth (h) at the small size as defined for the lowest treating layer, so long as the layer is evaluated in the non-perpendicular zones. See **Figure 6** in which "h" is illustrated to the equation (equation with all porosities, including three equations) of overburden pressure and water block, and it does figure out that the h's have a relationship with the type of application. Therefore, in the first layer, it requires to determine the whole overlying column which exerts a pressure on the lowest layer we are about to treat in the processes before or while producing. In calculations, the h is considered the thickness of the layer which has a cylinder-shaped geometric figure so as to cover a fully horizontal zone. This horizontal zone, in which the fluid pressure and the sudden gravitational forces emerged in the entire area of desired treating/drilling layer associated to the thickness of h through which movements and slips are made, could be led to compacting pores and displacing particles in the layers. These compacting and replacing occurrences can increase in the salty layers, and eventually unsteady the overburden pressure of the large contact area of the same layer and the boundary layers that slow the fluid moving forward or moving tools toward the boundary reservoir. Approaching these pressures to each other, it might damage to any moving tools in horizontal zone. Generally, in calculations of the

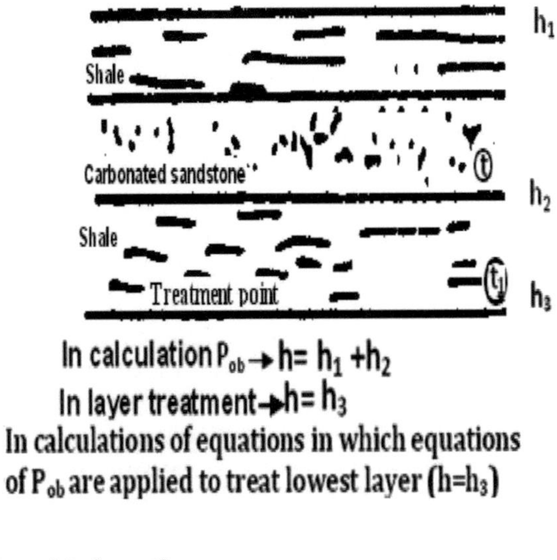


**Figure 6.**
*The total trend of h in $P_{ob}$ and $B_k$ [33, 36].*

underground operations such as water block in horizontal operations, we consider the whole thickness of the lowest layer as h. The decision to design a geometric section for the area of the well ($A_w$) in equations of $B_k$ depends on the desirable selected area around the wellbore. For this aim, as the part of the well/layer design, it is worthwhile determining the definite layer area using the integral techniques or other practical ways in industry and how $A_w$ is defined which should also be related to h. Even though an integral method provides the additional assurances to an effective area of layers without geometric shape, it may require a lot of time, and as such it may sometimes not practical. If we consider a well in which the zone to the length of L and the thickness of $h_2$ is treated to propel the fluid onward and or take away the obstacles around the wellbore for flowing the water injection issues (issues emanated from the scale, water block and phase trap, or any other arbitrary and tentative process of fluids and rocks that are caused to mechanical anisotropy and completion skins), then the addition of length has a relative relationship with the overburden pressure and consequently with damage caused by water block. And, the damages of rock and fluid are more in these long zones; thus, these high-angle wells have their own complexities while treating them to lessen the damages. Drawdown caused by these damages in the more consolidated formations is usually higher and against the plugged debris is more in the pore throats.

For more discussion on the variable L in equation of overburden pressure that is out of scope of this investigation, we suppose the damage is made in well to a large extent while drilling, and then the first L at the beginning of drilling is bit-length. After that the drill string length is added to the length of bit, this length is corresponding to the L in the overburden pressure equation that further affects on the mud, pressure drop near the wellbore, reservoir pressure, and any other treatment process in which open or casing-completed horizon is appropriate in increasing vertical permeability. Of course, these permeability increments decrease as the ratio of thickness to horizon length increases (see h in the $I_k$ and L in the $P_{ob}$ that have a direct and reverse relation with damage removed, respectively). For instance, in a slant well, usually more than 20°, this length is calculable using well angle and thickness drilled.

A summary of the main points on these equations in this capture of book is expressed, and it requires for an engineering to take some kind of action:

- The flow rate increases in permeable rocks which has a significant role in pushing the damage like water block, and performance of low-viscosity solvents is at a maximum amount in loose water-wet rocks. The efficiency of high-viscosity solvents and low viscous solvents in loose rocks is averagely more than that in dense rocks.

- As the pores of the rock saturate with water owing to the water-based fluid invasion, therefore, the residual oil phase would be pushed along with fractures because of the rate, pressure drop, and high capillary pressures. Furthermore, since the aperture size in faults is wide (mostly FW > 0.35 m), in equations of $B_k$ for salt water disposal wells (SWDW) is assumed the front of fluid injected into well during flowing through the area around the wellbore are not encountered to the fault.

- The decrease of porosity ($\varphi$) with depth (h) in the reservoirs is mostly true. But the $C_m$ or the same percentage solvent used according to the factors indicated in equations reduces over these variations. In such a situation, which the acid expending ($I_k$) estimated high, the rate of blocking ($q_B$) in such a low porosity media with relatively usual permeability becomes low, and subsequently the amount of the ($B_k$) is less. For this reason, the certainty of ($I_k$) as a measure to ascertain water block in equations of the ($B_k$) could help to effectively decide for treating the damage raised of water block in the zone around the wellbore that is surrounded by the other layers.

## Nomenclature

### Physical quantities

K = absolute permeability, md
$\rho$ = density, kg/$m^3$
V = volume, $m^3$
W = weight, kg
L = length, m
T = temperature, °F
$\mu$ = viscosity, kg/m-s
$A_w$ = sectional area, $m^2$
P = pressure, bar or psi
$\varphi$ = porosity, %
$\Delta P$ = pressure loss, psi
q = rate, $m^3$/s
$q_{mam}$ = maximum rate, $m^3$/s
$q_{mim}$ = minimum rate, $m^3$/s
$q_B$ = rate of blocking the fluid, $m^3$/min
$C_{mu}$ = mutual solvent concentration
$D_s$ = salinity degree, ppm
$H_{Mg}$ = hardness of Mg, ppm
pH = potential of hydrogen
$d_1$ = intergranular space, m
$d_2$ = inter-fracture space, m
$d_3$ = width fracture, m
$I_k$ = coefficient of Karimi (acid expending ability), mK or T°m
mk = a unit of measure of acid expending ability
$B_k$ = removed water block or power of damage removal by chemical, $m^3$/min $\times$ $(T°m)^{-0.5}$

### Subscripts

f = fluid
B = blocking of water
W = weight
B = symbol of blocking of water
r = rock
b = bulk
d = dry
w = water
w.w = water-wet
o = oil
o.w = oil-wet
ac = acid
mu = mutual
p = pore
ob = overburden pressure

HVMS = high-viscosity mutual solvent
LVMS = low-viscosity mutual solvent
HVSs = high-viscosity solvents
LVSs = low-viscosity solvents

**Superscripts**

$C_1$ = solvent percent numbers
A = convert coefficient for $k_{gf}$ to bar functions, etc.
$\mu_O$ = equation of oil viscosity
$P_{ob}$ = equation of overburden pressure
$B_k$ = removed water block equation, $\frac{m^3}{\min\sqrt{T^0 m}}$
$(B_k)_{o.w}$ = equation of removed water block in oil-wet oil wells
$(B_k)_{w.w}$ = equation of removed water block in water-wet oil wells

## Author details

Mohammad Karimi[1*], Mohammad Reza Adelzadeh[2], Mojtaba Mosleh Tehrani[3], Maryam Mohammadipour[4], Ruhangiz Mohammadian[5] and Abbas Helalizade[6]

1 Production and Drilling, Omidiyeh University, Khuzestan, Iran

2 Reservoir Engineering, NISOC, Khuzestan, Iran

3 Research and Development Center, NISOC (National Iranian South Oil Company), Khuzestan, Iran

4 Production Engineering, NISOC, Khuzestan, Iran

5 Geology, NISOC, Khuzestan, Iran

6 Reservoir, University of Petroleum Technology, Khuzestan, Iran

*Address all correspondence to: karimioilm@gmail.com

## References


[1] Mahadevan J, Sharma MM. Clean-up of water blocks in low permeability formations. In: Annual Technical Conference and Exhibition held in Denver, Colorado; October 5, 2003. SPE84216

[2] Mahadevan J, Sharma MM. Factors affecting cleanup of water blocks: A laboratory investigation. SPE Journal. 2005;**10**(3):238-246. SPE-84216-pa

[3] You L. The research on dynamics of aqueous phase trapping in tight sandstone gas reservoirs [dissertation]. Chengdu, China: Southwest Petroleum University; 2006

[4] Karimi M, Adelzadeh MR. A formula to calculate the overburden pressure of oil layers in various significant reservoir parameters using the experimental, macroscopic and microscopic data. In: GEOINDO SP305, The International Conference on Geology, Geotechnology, and Mineral Resources; Thailand; 2011

[5] Ali HS, Al-Marhoun MA, Abu-Khamsin SA, Celik MS. The effect of overburden pressure on relative permeability. In: Fifth Conference of Society of Petroleum Engineering. Middle East in Manama, Bahrain; 1987. pp. 335-340

[6] Hurst A, Huse M, Cartwright JA, Duranti D. Sand injectivity in deep-water classic reservoirs: Are they there and do they matter? In: Atlas of Deep-Water Outcrops. Vol. 120. New York: Elsevier; 2007

[7] Jilani SZ, Menouar H, Al-Majed AA, Khan MA. Effect of overbalance pressure on formation damage. Journal of Petroleum Science and Engineering. 2002;**36**:97-109

[8] Hao Z, Yili K. Deformation theory and stress sensitivity of tight sandstones reservoirs. Journal of Natural Gas Geoscience. 2004;**15**(5):482-485

[9] Qian Z, Jianfen D, Lichun C, et al. Study on the stress-sensitivity of the reservoir rock in sulige gas field. Journal of Drilling Fluid & Completion Fluid. 2006;**5**(23):29-30

[10] Allen JRL. Sedimentary structures: their character and physical basis. Developments in Sedimentology. 1984; **30**:593

[11] Compton RR. Geology in the Field. New York: John Wiley & Sons; 1985

[12] Matthews W, Kelly J. How to predict formation pressure and fracture gradient. Oil & Gas Journal. 1967;**65**(7): 92-106

[13] Karimi M, Adelzadeh MR, Mohammadypour M. Formula of definite point overburden pressure of reservoir layers. Egyptian Journal of Petroleum. 2014;**23**:175-182

[14] Karimi M, Helalizadeh A, Tehrani MM. Equations of overburden pressure a core sample at laboratory. International Journal of Petroleum Engineering. 2014;**2**(1):143-156

[15] Bennion DB, Thomas FB, Bietz RF, et al. Water and hydrocarbon phase trapping in porous media diagnosis, prevention and treatment. Journal of Canadian Petroleum Technology. 1996; **10**(35):29-36

[16] Bennion DB, Thomas FB, Schulmeister B, Romanova UG. Water and oil base fluid retention in low permeability porous media—An update. In: Paper presented at the Petroleum Society's 7th Canadian International Petroleum Conference; Canada; 2006

[17] Çoşkuner G. Completion operations in low permeability deep basin gas

reservoirs: To use or not to use aqueous fluids, that is the question. Journal of Canadian Petroleum Technology. 2006; **10**(43):24-28

[18] Mirzaei A, Moghadasi J, Masihi M, Dalvand K, Oraki K. A new formula for evaluation of severity of aqueous phase trapping in oil reservoirs. In: The 1st International Applied Geological Congress. Mashhad, Iran: Islamic Azad University; 2010. pp. 26-28

[19] Mahadevan J. Novel Fluids for Productivity Enhancement in Tight Gas Formations. University of Tulsa: Report sponsored by Research Partnership to Secure Energy for America (RPSEA); 2011. p. 15. No. 07122-36

[20] Abrams A, Vinegar HJ. Impairment mechanisms in Vicksburg tight gas sands. In: Paper SPE/DOE 13883 presented at the SPE/DOE Joint Low Permeability Gas Reservoirs Symposium held in Denver, Colorado; 1985. pp. 19-22

[21] Holditch AS. Factors affecting water blocking and gas flow from hydraulically fractured gas wells. Journal of Petroleum Technology. In: SPE-AIME 53nd Annual Conference and Exhibition, Houston; 1979. pp. 1515-1524

[22] Lowrie W. Fundamentals of Geophysics. New York: Cambridge University; 1997

[23] Leone AJ, Scott ME. Characterization and control of formation damage during water flooding of a high-clay-content reservoir. SPE Reservoir Engineering. 1988;**4**(3):1279-1286. SPE 16234

[24] Civan F. A multi-purpose formation damage model. In: Formation Damage Control Symposium; 14–15 February; Lafayette, Louisiana; 1996. SPE31101-MS

[25] Kamath J, Catherine L. Laboratory based evaluation of gas well deliverability loss due to water blocking. In: Paper SPE 63161 presented at the SPE Annual Technical Conference and Exhibition held in Dallas, Texas; October 1–4, 2000

[26] Fahes M, Firoozabadi A. Wettability alteration to intermediate gas-wetting in gas-condensate reservoirs at high temperatures. In: Paper SPE 96184 was first presented at the 2005, SPE Annual Technical Conference and Exhibition, Dallas, and revised for publication in SPE Journal; 2005

[27] Parekh B. Cleanup of water blocks in depleted low-permeability reservoirs [Master's thesis]. Austin, TX: University of Texas at Austin; 2003

[28] Lake LW. Enhanced Oil Recovery. New Jersey: Prentice Hall; 1989

[29] Abramowitz M, Stegun IA. Handbook of Mathematical Functions Mechanics. Washington, D.C.: Tenth Printing; 1972. pp. 9-65

[30] Chen Z, Huan G, Ma Y. Computational methods for multiphase flows in porous media. Computational Science & Engineering. 2006. pp. 399-443. ISBN: 978-0-89871-606-1

[31] Gray WG. On the definition and derivatives of macroscale energy for the description of multiphase systems. Advances in Water Resources. 2002;**25** (8–12):1091-1104

[32] Gray WG, Miller CT. Thermodynamically constrained averaging theory approach for modeling flow and transport phenomena in porous medium systems: 2. Foundation. Advances in Water Resources. 2005;**28**: 181-202

[33] Karimi M, Adelzadeh MR, Helalizadeh A. Equations of Water Block: An Injection Process for Removal

of Damage in Oil Wells. International Journal of Modeling and Simulation for the Petroleum Industry of Brazil. 2016; **10**(1):31-44

[34] Lee WJ, Rollins JB, Spivey JP. Pressure Transient Testing. SPE Textbook Series. Vol. 9. Richardson, Academic Publisher of Scientific Research, TX; 2003

[35] Efendiev Y, Hou TY. Multiscale Finite Element Methods. Surveys and Tutorials in the Applied Mathematical Sciences. Vol. 4. New York: Springer Verlag; 2009

[36] Karimi M, Adelzadeh MR, Helalizadeh A. New equations of water block in oil wells. Universal Journal of Petroleum. 2016;**4**:19-39

[37] Karimi M, Adelzadeh MR, Helalizadeh A. Equations to calculate the overburden pressure in water layers of definite dimensions using experimental, macroscopic and microscopic data. Journal of Petroleum Science and Engineering. 2013;**101**(4): 68-74

[38] Crabtree M, Eslinger D, Fletcher P, Miller M, Johnson A, King G. Fighting scale—Removal and prevention. Oilfield Review. 1999;**Autumn**:30-45

[39] Eakin JL, Miller JS, Eckard WE. Removal of water block from gas-producing formations. In: Paper presented at the Spring Meeting of the Eastern District, API Division of Production; 1965. pp. 204-217

[40] Jordan JR, Shirley OJ. Application of drilling performance data to overpressure detection. JPT. 1966

[41] Tearpock DJ, Bischke RE. Applied Subsurface Geological Mapping. Englewood Cliffs, New Jersey: Prentice-Hall Inc; 1991. pp. 645-648

[42] Theys PP. Log Data Acquisition and Quality Control. 2nd ed. Paris: Editions Technip; 1999. p. 233-240; 310-320

[43] Taheri A, Zahedzadeh M, Masoudi R, Ataei A, Roayaei E, Fakhri H. Simulation and experimental studies of mineral scale formation effects on performance of Sirri-C oil field under water injection. Iranian journal of chemistry and chemical engineering. 2011;**3**:30

Chapter 7

# The Risk of Potential Cross Border Transport of Oil Spills in the Semi-Enclosed Eastern Mediterranean Sea

*Steve Brenner*


## Abstract

Environmental risks posed by oil spills in semi-enclosed basins are more pronounced than those in the open ocean due to potential deposition along long segments of the coastlines. As a semi-enclosed sea, the Mediterranean is highly vulnerable to pollution events. Recent discoveries of major oil and natural gas reserves in the eastern Levantine basin have led to accelerated drilling, with several countries at various stages of exploration and production and others having mapped blocks for licensing, thereby significantly increasing the risks of a potential spill. Due to drilling by multiple, adjacent countries, any spills from deep water wells will be prone to cross border transport due to the highly variable winds and ocean currents. This risk is assessed through a series of simulations with an oil spill model forced with high resolution ocean currents and winds. The scenarios considered are well blowouts of several weeks duration, located within the drilling zones of each of various countries. Models such as this provide the basis for further environmental assessment and risk analysis. They also emphasize the importance of multinational cooperation to respond to and mitigate the environmental impacts which would result from a potential oil spill from any of the countries involved.


**Keywords:** oil spill modeling, oil slick dispersion, cross border pollution transport, ocean model downscaling, eastern Mediterranean Sea gas and oil exploration

## 1. Introduction

Environmental risks posed by oil spills in semi-enclosed basins are more pronounced than those in the open ocean due to potential deposition along long segments of the coastlines. As a prototype semi-enclosed sea, the Mediterranean is sensitive and vulnerable to pollution events in general, and specifically to potential oil spills from ships, offloading terminals, pipelines, or wells. As a semi-enclosed sub-basin of the Mediterranean Sea, the eastern Mediterranean can be expected to be even more vulnerable to pollution. During the past 10–15 years significant reserves of oil and natural gas have been discovered in the eastern Mediterranean Sea. The US Geological Survey [1] estimates that there are more than $3.45 \times 10^{12}$ $m^3$ of recoverable natural gas and $1.7 \times 10^9$ barrels of recoverable oil in these reserves, most of which is located beneath the seafloor of the eastern Levantine Basin. These reserves are located within the territorial waters and/or the exclusive economic zones (EEZ) of five countries—Egypt, Israel, Cyprus, Lebanon, and Syria. In principle each of the countries of the region has the rights to explore and drill within its EEZ. However due to the relatively small size of the eastern Levantine Basin, the different exploration zones sometimes abut one another, as shown in **Figure 1**. The conflicting claims of neighboring countries often lead to partial overlap of adjacent EEZs [2]. These disputes are further exacerbated by the ongoing political conflicts in the region. The small size of this semi-enclosed basin also leads to increased environmental vulnerability in the case of an accidental oil spill with a high potential risk for cross border pollution transport due to the prevailing winds and the near surface ocean currents [3, 4].

Studies of oil slick dispersion in other semi-enclosed basins and seas [5–8] have demonstrated the risk of widespread oil deposition from the slick along large segments of the coastlines. However in many cases cross border transport was not a consideration since the particular sea or basin was contained mostly within the territory of only one country [6–8]. Nevertheless, concerns regarding the attribution of responsibility and legal accountability for cross border or transboundary transport of marine pollution and its detrimental effects have been increasing over the past 20–30 years [9, 10].

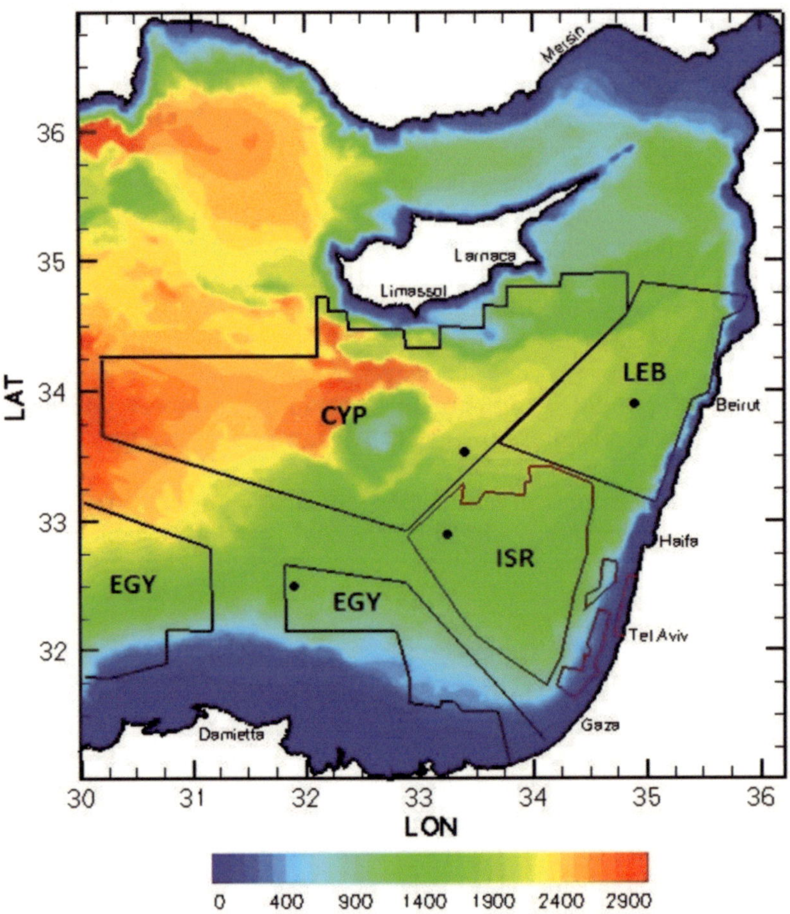


**Figure 1.**
*Bathymetric map of the eastern Levantine basin showing the approximate boundaries of the oil and natural gas exploration zones of Cyprus (CYP), Lebanon (LEB), Israel (ISR), and Egypt (EGY). Black dots indicate locations of the hypothetical discharge points.*

The primary goal of this study is to demonstrate the potential for cross border transport of an oil slick discharged from several hypothetical, offshore, deep water, drilling platforms located within the exploration zones in the EEZs of each of the adjacent countries that are actively exploring or exploiting the hydrocarbon energy reserves in the region. This is accomplished by running an oil spill dispersion model for a series of worst case scenarios consisting of a 30 day continuous discharge from a well blowout during various periods of representative meteorological and oceanographic conditions. These hypothetical simulations are not intended to represent any specific existing or planned well, but are designed to assess what could potentially occur in the regions of interest. Furthermore, the focus here is on deep water sites since shallow water wells in this region will be much closer to the coast (~10–20 km or less) and slicks from the latter will most likely reach the coast in close proximity to the well within a few days and will therefore be less prone to cross border transport [4, 11]. While blowouts may be rare events, they tend to contribute a disproportionately large percentage of the total spill volume from all sources. For example, over a 39 year period (1969–2007), well blowouts accounted for less than 1.5% of the offshore oil spill incidents in the United States but contributed nearly 85% of the total volume spilled [12]. The results of this study are qualitative in the sense that the level of risk is not numerically ranked or scored, but rather it is considered sufficient to show that a slick enters the EEZ or reaches the shores of a country other than the one responsible for the spill. In this respect, for the scenarios considered there is a very high probability of cross border transport of the slick. This emphasizes the shared multinational responsibility to control and prevent catastrophic pollution events in this environmentally sensitive and vulnerable region.

## 2. The atmospheric and oceanographic setting

The Mediterranean Sea is often considered the prototype semi-enclosed basin driven by net evaporation and therefore producing relatively dense, saline water which flows out into the adjacent ocean where it quickly sinks. It is connected to the North Atlantic Ocean through the narrow and relatively shallow (~270 m) Straits of Gibraltar. It is divided into the western and eastern sub-basins at the Straits of Sicily with the eastern basin accounting for nearly 2/3 of the surface area of the Mediterranean. The residence time is on the order of 100 years which makes this basin sensitive and vulnerable to major pollution events. The sea is located between the mid-latitude Westerlies to the north and the hot, dry subtropical deserts to the south and is characterized by the unique Mediterranean type climate consisting of cool, wet winters and hot, dry summers. During winter the weather over the Levantine basin is affected mainly by migratory cyclones, many of which originate in the Gulf of Genoa, near Crete or near Cyprus [13]. During summer the eastern Mediterranean is mostly under the influence of a westward extension of the Asian thermal low, leading to a strongly persistent pattern referred to as the Persian Trough [14]. Consequently the predominant winds over the Levantine basin throughout most of the year have a strong westerly component and blow mostly from the northwesterly to southwesterly sector. Monthly mean winds for December 2010 and September 2007 are shown in **Figures 2** and **3**, respectively.

The apparently larger mean wind vector in September, as compared to December, is due to the strong persistence of the wind direction in this season. However, in December the mean wind speeds over the sea are typically 20–40% stronger than in September, while the standard deviations are 50–100% larger in December as compared to September. This is due to the relatively stronger and

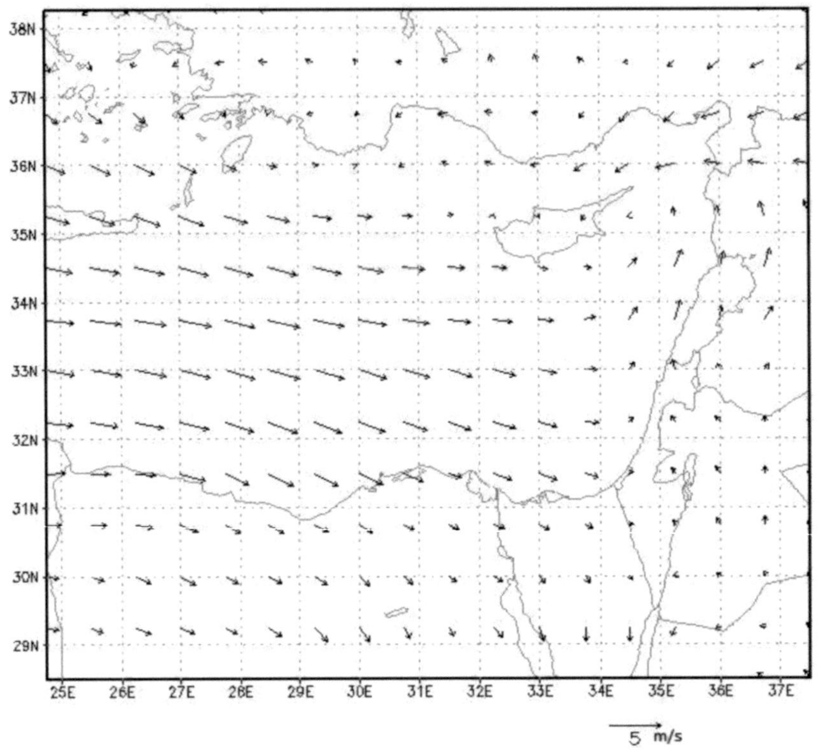


**Figure 2.**
*Monthly mean 10 m winds for Dec 2010.*

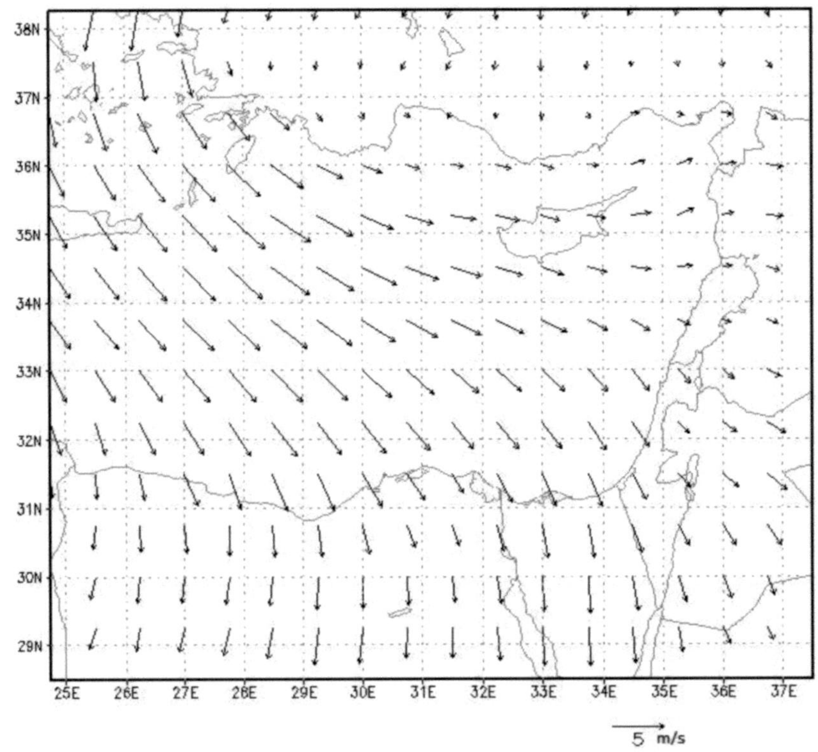


**Figure 3.**
*Monthly mean 10 m winds for Sep 2007.*

variable direction winds in the winter associated with the migratory cyclones that pass through this region. These seasonal wind characteristics have important implications for the spreading of oil slicks as will be seen below in Section 4.

The annual evaporation from the Mediterranean exceeds the fresh water input by precipitation and runoff with a net loss of ~0.64 m $y^{-1}$ [15]. This leads to a long term, mean anti-estuarine thermohaline circulation pattern in which less saline water flows in from the ocean in the upper layer and is nearly balanced by the subsurface outflow of saline water produced in the basin. Most of this saline outflow consists of Levantine Intermediate Water (LIW) which is formed at various locations in the Levantine Basin [16]. A variety of dynamical circulation features, covering a range of temporal and spatial scales, is superimposed on this thermohaline cell, including convective deep water formation, sub-basin scale gyres, meandering jets, and a highly energetic mesoscale eddy field [17]. One of the most prominent features in the Levantine Basin is the meandering Mid-Mediterranean Jet (MMJ) which flows eastward through the center of the basin where it flanks various quasi-permanent or recurrent sub-basin scale features such as the cyclonic Rhodes gyre and the anticyclonic Mersah Matruh and Shikmona gyres [16, 18]. The MMJ bifurcates off the coast of Egypt where it splits with a southern branch that forms the cyclonic shelf break jet which then turns northward flowing along the coasts of Israel, Lebanon, and Syria [19], followed by a turn to the west flowing along the coast of Turkey. Examples of the winter and summer upper layer (30 m), monthly mean currents for the eastern Levantine Basin are shown for Dec 2010 in **Figure 4** and for Aug 2008 in **Figure 5**. These current patterns together with the prevailing

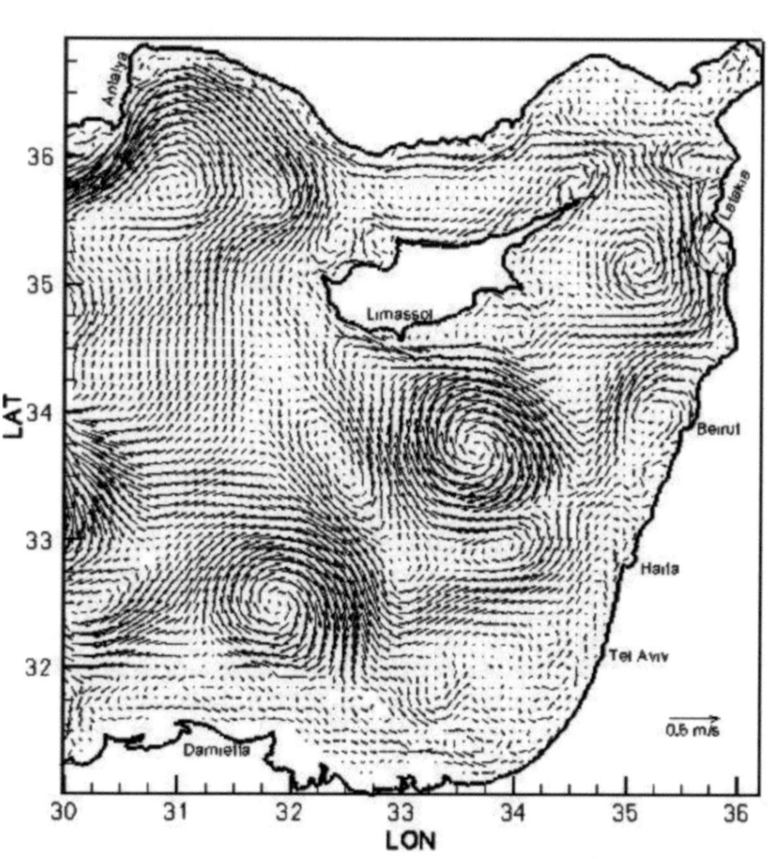


**Figure 4.**
*Upper layer (30 m) monthly mean currents for Dec 2010.*

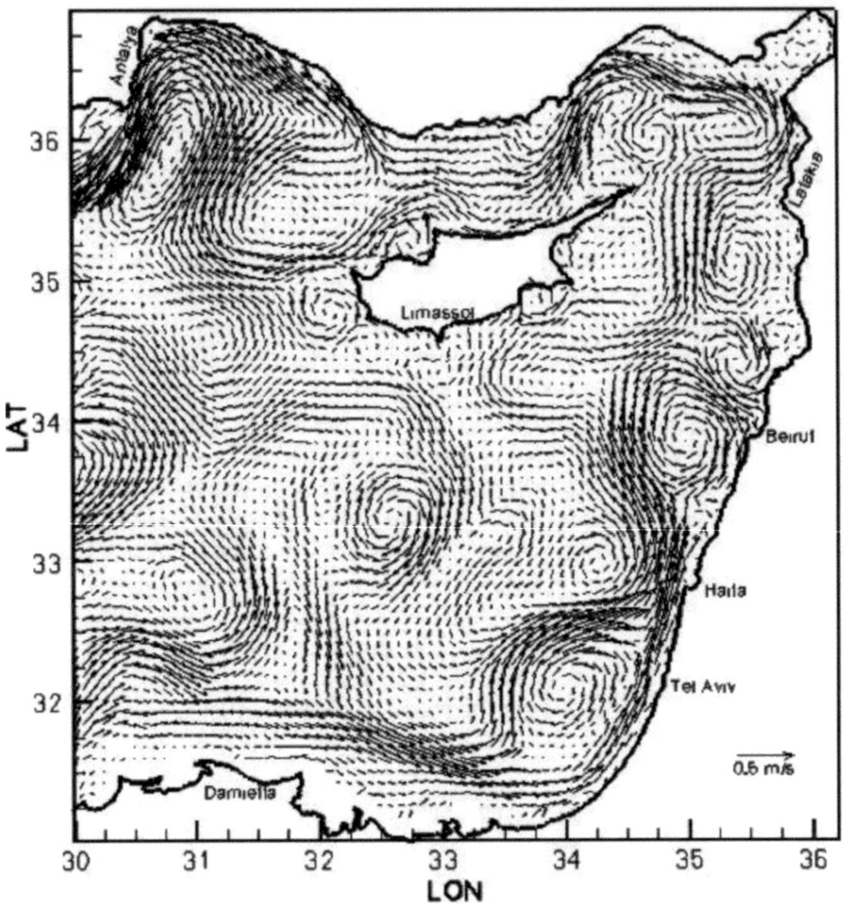


**Figure 5.**
*Upper layer (30 m) monthly mean currents for Aug 2008.*

winds are expected have a major impact on the dispersion of near surface oil slicks released from offshore locations in this region.

## 3. Methodology

In order to assess the risk of cross border transport of an oil slick discharged from hypothetical offshore, deep water wells located in the EEZs of each the four countries actively involved in exploration or production, a widely used oil spill dispersion model has been run for a variety of discharge scenarios during different, representative meteo-oceanographic conditions. As noted in the introduction, all of these scenarios assume a continuous 30 day discharge of oil from an uncontrolled well blowout with no preventative or mitigating measures such as deployment of booms or dispersants implemented. In this respect the simulation conducted can be considered worst case scenarios. A 30 day period also ensures that the slick will spread significantly and will be subjected to a wide range of winds and currents. The oil spill model requires winds and near surface ocean currents to compute the trajectory and dispersion of the slick. It is therefore run in two steps. First a high resolution, ocean circulation model is run to downscale the currents from an ocean reanalysis data set. In the second step, these currents and the winds are used as input to the oil spill fate model which computes the trajectory of the slick as well as various weathering processes.

### 3.1 Ocean circulation model

The currents used to drive the oil spill model are generated using an expanded domain and higher resolution version of the model developed as part of an operational ocean forecasting system for the southeastern Mediterranean Sea [20]. It is based on the Princeton Ocean Model (POM) [21] which is a three dimensional, free surface, time dependent primitive equations model. POM uses an Arakawa-C grid in the horizontal and a terrain following, sigma coordinate in the vertical. It contains full thermodynamics and includes a higher order turbulence closure scheme to account for sub-grid scale vertical mixing. It is forced at the surface with climatological heat and fresh water fluxes [15] corrected with a Newtonian relaxation or nudging to the daily varying surface temperature and salinity extracted from the Mediterranean Sea reanalysis [22]. Six hourly winds were extracted from the ECMWF Interim Reanalysis [23]. Initial conditions and lateral boundary conditions at the open western boundary are derived from the daily varying, three dimensional Mediterranean Sea reanalysis [22]. The horizontal resolution of the model is 1/60° in latitude and longitude (1.85 × 1.55 km) and there are 30 unevenly spaced sigma levels in the vertical. The domain and bathymetry used in the model are shown in **Figure 1**. The examples of the near surface currents shown in **Figures 2** and **3** were produced by this model.

### 3.2 Oil spill model

The oil spill fate model used to assess the dispersion of slicks discharged from the hypothetical deep water wells is MEDLSIK [24, 25]. In the model, the surface oil slick is advected and dispersed by the direct action of the near surface currents and the wind forcing. The slick is assumed to be composed of a very large number (tens or hundreds of thousands) of particles which are transported following a Lagrangian trajectory determined by the incremental displacement due to the currents, the winds, and small scale, horizontal turbulent diffusion based on the random walk hypothesis. The model computes the total oil budget with a full accounting of oil that remains on the surface, oil that is mixed in the water column, and oil that is beached. It also accounts for weathering of the oil through physio-chemical processes such as evaporation, emulsification, and small scale, vertical turbulent mixing [24, 25]. MEDSLIK has been tested extensively and widely used in many locations around the Mediterranean [3, 4, 11, 26, 27] as well as in other regions of the world [6].

### 3.3 Scenarios

As noted above, the goal of this study is to demonstrate the vulnerability of this semi-enclosed basin to the risk of cross border transport of an oil slick originating from spills at hypothetical, yet representative, offshore, deep water wells potentially operated by four different countries in the region. To this end, worst case oil spill scenarios are considered, consisting of uncontrolled discharges from continuous, 30 day well blowouts at each of the four locations indicated by the black dots in **Figure 1**, with no preventative or mitigating measures implemented. Selection of the meteo-oceanographic conditions for the simulations was done following the guidelines of the Israel Ministry of Environmental Protection for conducting environmental impact studies of offshore gas exploration or production. They have specified four representative 30-day periods including: (1) at least one extreme winter storm (Dec 2010–Jan 2011), (2) a winter period with at least two typical storms (Jan–Feb 2008), (3) a typical summer period with persistent northwesterly winds (Jul–Aug 2008), and (4) a transition season (late summer—autumn) period with several episodes of strong easterly winds (Sep–Oct 2007). The oil from each blowout is assumed to be medium grade with an API gravity of 33° (within the range of exploratory discoveries in this region [3]) and discharged at a rate of 2000 barrels per day (total of 60,000 barrels from each blowout).

## 4. Results and discussion

An overview of the results of all 16 simulations is presented in **Table 1** where the oil budget at the end of 30 days is broken down by percentages of the total discharged, into the amounts evaporated, remaining on the surface, dispersed (vertically mixed) in the water column, and amount deposited on the coast. The next to last column of the table lists the coasts of the countries affected, while the last column gives the time (in days) until the first beaching of oil.

Perhaps the most important result from **Table 1** is that in 88% of the cases (14 out of 16) cases, the coastlines of two or more countries will be affected, while in 25% of the cases the coasts of four or more countries will be impacted. In two cases the impact is limited to the coast of only one country. In one of those cases

| Period | Source | Evap | Surf | Disp | Depo | Coasts affected | Beaching (days) |
|---|---|---|---|---|---|---|---|
| 12/10–01/11 | C | 42.1 | 43.2 | 14.0 | 0.7 | **C**, I, **L**, S | 21 |
| | L | 42.1 | 36.9 | 12.3 | 8.6 | C, I, **L**, **S**, T | 4 |
| | I | 42.06 | 41.4 | 16.5 | 0.04 | C | 26 |
| | E | 42.1 | 38.4 | 16.8 | 2.5 | **E**, I | 21 |
| 01/08–02/08 | C | 42.1 | 34.6 | 17.1 | 6.1 | **I**, L | 19 |
| | L | 42.1 | 32.2 | 15.3 | 10.0 | C, L, **S** | 17 |
| | I | 42.1 | 31.0 | 16.4 | 10.1 | E, **I**, L | 16 |
| | E | 42.1 | 30.6 | 15.9 | 11.0 | **E**, I | 11 |
| 07/08–08/08 | C | 42.1 | 35.8 | 14.5 | 7.6 | C, L, **S**, T | 18 |
| | L | 42.1 | 15.1 | 7.1 | 35.6 | **L**, **S** | 6 |
| | I | 42.1 | 35.0 | 14.9 | 8.0 | L, **S** | 16 |
| | E | 42.1 | 34.6 | 14.4 | 9.0 | **I**, L | 19 |
| 09/07–10/07 | C | 42.1 | 40.3 | 14.2 | 3.4 | E, **I** | 18 |
| | L | 42.0 | 27.0 | 6.3 | 24.5 | C, **L**, I, S | 13 |
| | I | 42.1 | 22.1 | 11.5 | 24.1 | E, **I** | 14 |
| | E | 42.1 | 23.1 | 12.3 | 22.0 | **E** | 19 |

*Coastlines affected are also listed according to C, Cyprus; E, Egypt; I, Israel; L, Lebanon; S, Syria; T, Turkey. The most severely impacted coastlines are indicated by bold letters. The last column gives the number of days until first beaching of the oil.*

**Table 1.**
*Components of oil budget (in%) after 30 days—evaporated, remaining on surface, vertically dispersed, and deposited on the coast.*

practically no oil reaches any cost (only a negligible amount on the southern coast of Cyprus) while in the other case, the coastline affected belongs to the same country in whose domain the spill originated (Egypt). Once again it should be noted that in this study a coastline is considered to be affected as long as some oil is deposited. This measure is strictly qualitative with no numerical ranking or risk factor computed. The distribution of the oil on the coast is not uniform and can be very patchy. In many cases the most severely impacted zones (i.e., highest concentrations) are limited to one or several relatively short stretches of coastline. In all of the cases considered, slightly more than 42% of the discharged oil evaporates. This is due primarily to the API gravity of the oil. As a general rule of thumb, for heavier oil, a smaller portion would evaporate while for lighter oil more of the oil will evaporate. For example, several scenarios were rerun with 40° API (a value that is representative of condensate that could potentially leak from a natural gas well). In these cases slightly more than 53% of the oil evaporated with the other components of the oil budget reduced accordingly, but the spatial distributions of the oil remaining on the surface and deposited on the coast were similar to the original simulations with 33° API gravity.

In 14 of the 16 cases, the second leading term in the oil budget, after evaporation, is the oil remaining on the surface. The percentage ranges between 23.1 and 43.2% with an average of 32.6%. In the remaining two case, the amounts are 15.1–22.1%. The next term in the budget, in order of magnitude, is oil dispersed (vertically mixed) in the water column with an average of 13.7%. These two components of the budget, which represent the portion of oil that remains in the water, account for most of the oil that remains after evaporation. The last term in the oil budget is the amount deposited on the coast, which ranges from 0.04 to 35.6% with an average of 11.5% of the total amount discharged. In two of the cases (both in Dec 2010), the amount deposited on the coast was less than 1%. In four of the cases more than 22% of the oil is deposited on the coast. Three of these four cases occurred in Sep 2007 and can be attributed to the persistent northwesterly winds in the southeastern part of the basin and westerly winds in the northeastern part of the basin (**Figure 3**).

The last column in **Table 1** shows the time in days until the first oil is deposited on the coast. This time ranges from 4 to 26 days with an average of 16 days. From the table it is noticeable that the hypothetical slick originating from the EEZ of Lebanon will tend to reach the coast faster than the slicks originating from other EEZs. This is due to the closer proximity of this well to the coast and the relatively confined domain, as compared to the others, combined with the predominant westerly component of the winds in this region. The other interesting point to note in the table is that the amount of oil reaching the coast is consistently lowest in the Dec 2010 simulations. The strong near surface currents, dominated by two large and intense anticyclonic features (see **Figure 4**), combined with the relatively strong and variable winds during this period lead to widespread dispersion of the slicks across the basin with most of the unevaporated oil remaining in the sea as shown in **Figure 6**.

To better appreciate the important roles of near surface currents and winds in spreading the slick, the next four figures present the spatial distribution of the oil remaining on the surface and the oil deposited on the coast at the end of the simulations from Dec 2010 and Sep 2007. **Figure 6** shows the oil remaining on the sea surface at the end of the four hypothetical well blowouts in Dec 2010. This period was chosen since the amount of oil that was deposited on the coast was minimal while the dispersion of the surface oil was most widespread. This period was characterized by an extreme winter storm towards the beginning and strong, but highly variable direction, winds throughout the period. The general tendency for

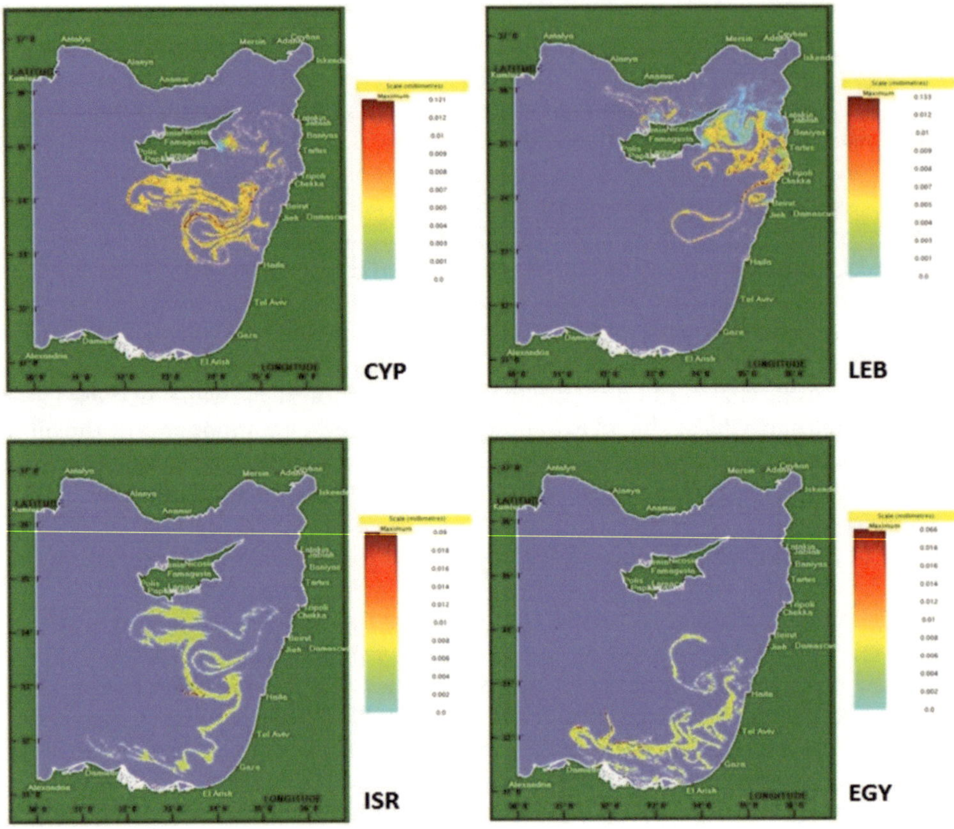


**Figure 6.**
*Oil remaining on the surface at the end of the four hypothetical 30 day well blowout simulations in Dec 2010.*

widespread dispersion in these cases can be explained by the relatively strong and variable winds. However the details of the spreading in streamers and filaments are clearly a result of the mesoscale variability of the near surface currents.

**Figure 7** shows the oil deposited on the coast for the Dec 2010 simulations. The slick originating from the Cyprus and Lebanon EEZs are the most likely to result in cross border transport and lead to deposition on the coastline of four different countries. The slick originating from the Egyptian EEZ will also reach the coast but almost exclusively affecting the Egyptian coast. The exception here is the slick originating in the EEZ of Israel which almost completely remains at sea, with a negligible amount of oil (only 0.04%) deposited on the southern coast of Cyprus.

For comparison, **Figures 8** and **9** show the results for the Sep 2007 simulations. This period was chosen to contrast with Dec 2010 since on average here the percentage of coastal deposition was significantly larger than for the other cases. In this case the transport of the slick was strongly controlled by the combined effects of the persistent winds and the mesoscale features of the currents. The slicks originating from the EEZs of Cyprus, Egypt, and Israel are all transported southward in the open sea by the strong northwesterly winds. It is interesting to note that the transport is highly focused in thin filaments following the near surface currents. When the slick reaches the continental shelf it is transported alongshore by the cyclonic, shore parallel current system [19]. The behavior of the slick originating the EEZ of Lebanon is somewhat different. In this region the winds have a stronger westerly component which transports the slick rapidly towards the coast. Upon reaching the continental margin, the oil is then transported northward by the combined action of the shore parallel jet and a series of energetic, northward moving mesoscale

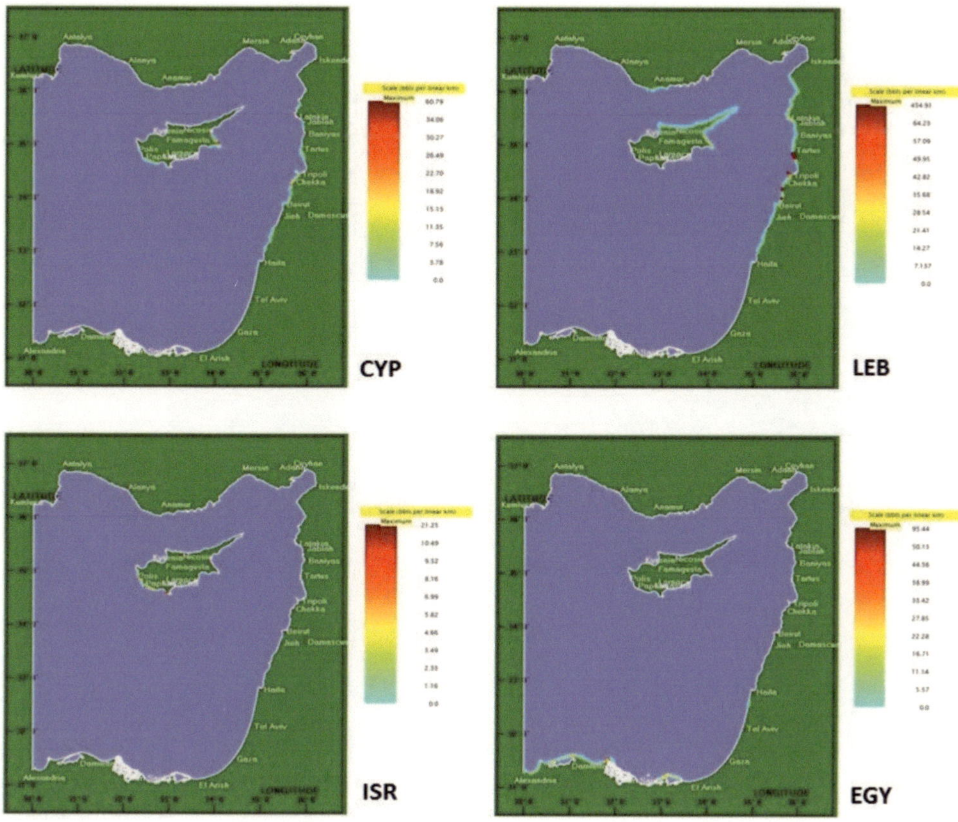


**Figure 7.**
*Oil deposited on the coast at the end of the four hypothetical 30 day well blowout simulations in Dec 2010.*

eddies. Consequently the shape of the slick is more diffuse and less focused than the other three slicks.

The oil deposited on the coast from the four slicks in Sep 2007 is shown in **Figure 9**. In all four cases it is clear that the pattern of coastal deposition is strongly controlled by the shore parallel current that flows over the continental shelf [19]. Consequently the oil primarily affects the coasts of Egypt, Israel, Lebanon and Syria. Only a small amount of oil from the Lebanon slick reaches the north coast of Cyprus.

The main goal of this study has been to provide a broad overview of the potential risk from hypothetical oil spills originating in the EEZ's of four different countries in the region that are at various stages of exploration and production of natural gas and oil. Other modeling studies of oil spills in this region [3, 4, 11] have focused on spills originating from the EEZs of only a single country. This study also differs from the others in terms of the longer duration of the spills and simulations, as well as the focus on spills originating from deep water platforms which tend to be further from the coast. Nevertheless, even in those studies there was often a tendency to see some cross border transport of the slicks, at least to the coasts of the immediately adjacent countries. Based studies like this it is clear that this region, in which gas and oil exploration and production has proceeded at an accelerated rate, is highly susceptible and vulnerable to cross border transport of oil slicks and the resulting environmental damage. None of countries of this region are immune to the risk. They all share the vulnerability and must also accept and share the responsibility. The need for action and cooperation to control and respond to cross border marine pollution events was already recognized more than 2 decades

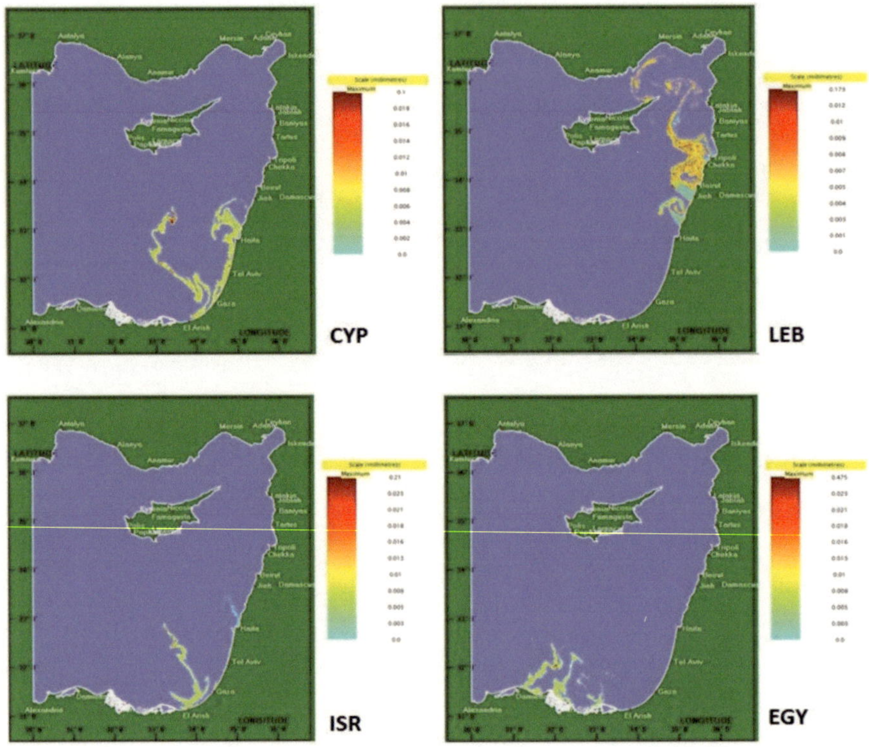


**Figure 8.**
*Oil remaining on the surface at the end of the four hypothetical 30 day well blowout simulations in Sep 2007.*

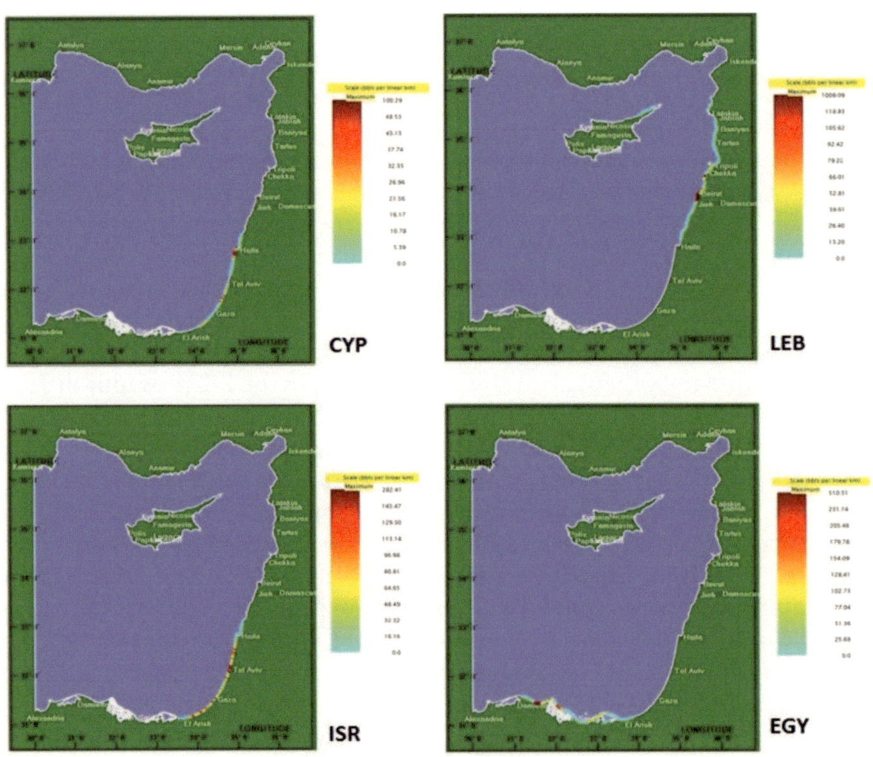


**Figure 9.**
*Oil deposited on the coast at the end of the four hypothetical 30 day well blowout simulations in Sep 2007.*

ago when a trilateral agreement establishing a sub-regional contingency plan was signed between Cyprus, Egypt, and Israel in 1995 [28]. More recently, an additional sub-regional contingency plan agreement was signed between Cyprus, Greece, and Israel last year [29]. Both of these agreements are important and encouraging and hopefully additional sub-agreements or parallel (bilateral) agreements will be signed that will allow for crucial cooperation in response to major oil spills despite the ongoing political tensions in the region.

## 5. Conclusions

In general, the environmental risks posed by oil spills in semi-enclosed sea, basins or bays are more pronounced than those in the open ocean due to potential deposition along long segments of the coastlines. As a prototype semi-enclosed sea, the Mediterranean is sensitive and vulnerable to pollution events in general, and specifically to potential oil spills from ships, offloading terminals, pipelines, or wells. As a semi-enclosed sub-basin of the Mediterranean Sea, the eastern Mediterranean can be expected to be even more vulnerable to pollution and therefore serve as a test case for investigating the risks associated with widespread dispersion and beaching of oil slicks. Recent discoveries of extensive hydrocarbon energy reserves beneath the seafloor of the eastern part of the Levantine basin have led to accelerated gas and oil exploration by four countries in the region—Cyprus, Egypt, Israel, and Lebanon. Two countries, Cyprus and Israel, have already begun production. In addition to the general risks posed by an oil spill in a restricted sea due to the patterns of the prevailing winds and sea surface currents, the active exploration and exploitation by multiple, adjacent countries make this region potentially sensitive and vulnerable to significant cross border transport. The risk of cross border transport is especially acute from spills originating in deep water platforms located in the open sea due the time required to reach the coast. Previous modeling studies of hypothetical oil spills in this region have generally focused on spills originating from the EEZ of only a single country. The goal of this study was to highlight the vulnerability of the eastern Levantine basin to the pronounced environmental risks of an oil spill in a semi-enclosed sea or basin combined with the unique situation of multiple, adjacent countries actively exploring and exploiting oil and natural gas reserves, thereby adding the consideration of risks of cross border transport. Thus a comparative overview is provided considering potential spills originating from deep water wells in the EEZs of four different countries. Due to the relatively large distance of these wells from the coast (tens of to more than 100 km) the oil can spread across large areas and will typically take 2–3 weeks to reach a coast. Consequently the risk for cross border transport of a slick originating from any of the EEZs is very real. Simulations were conducted for hypothetical 30 day, continuous, uncontrolled spills from four deep water platforms in different seasons. In 88% of the cases considered (14 out of 16) oil was deposited on the coast of at least one country other than the country responsible for the spill. On the other hand, the relatively long period until the oil reaches the coast is in principle important for allocating and deploying resources to contain the slick and to mitigate the damage. This emphasizes the importance of multinational cooperation in developing contingency and response plans and procedures in regions where several countries in close proximity are simultaneously producing oil or natural gas. It also highlights the importance of mutual responsibility to protect the marine environment since no country will be immune from potentially causing and subsequently suffering from the damaging effects of cross border pollution transport.

## Acknowledgements

We would like to thank Dr. George Zodiatis of the Cyprus Oceanographic Center for providing the oil spill model, MEDSLIK. Wind data were provided by the ECMWF and the oceanographic data were provided by the EU COPERNICUS Marine Environment Monitoring Service. This work was conducted as a downstream application and follow-on to several EU funded projects that supported the development of the Mediterranean Ocean Forecasting System (MFS).

## Conflict of interest

There are no known conflicts of interest related to this work.

## Author details

Steve Brenner
Department of Geography and Environment, Bar Ilan University, Ramat Gan, Israel

Address all correspondence to: sbrenner@mail.biu.ac.il

## References

[1] US Geological Survey. Assessment of undiscovered oil and gas resources of the Levant basin province, eastern Mediterranean. US Geological Survey, US Department of the Interior Fact Sheet 2010-3014; 2010. 4 pp

[2] Stocker J. No EEZ solution: The politics of oil and gas in the eastern Mediterranean. Middle East Journal. 2012;**66**(4):579-597. DOI: 10.3751/66.4.11

[3] Alves TM, Kokinou E, Zodiatis G, Lardner R, Panagiotakis C, Radhakrishnan H. Modeling of oil spills in confined marine basins: The case for early response in the eastern Mediterranean Sea. Environmental Pollution. 2015;**206**:390-399. DOI: 10.1016/j.envpol.2015.07.042

[4] Brenner S. Oil spill modeling in the southeastern Mediterranean Sea in support of accelerated offshore oil and gas exploration. Ocean Dynamics. 2015;**65**:1685-1697. DOI: 10.1007/s10236-015-0902-2

[5] Amir-Heidari P, Raie M. Probabilistic risk assessment of oil spill from offshore oil wells in Persian gulf. Marine Pollution Bulletin. 2018;**136**:291-299. DOI: 10.1016/j.marpolbul.2018.07.068

[6] Yu F, Xue S, Zhao Y, Chen G. Risk assessment of oil spills in the Chinese Bohai Sea for prevention and readiness. Marine Pollution Bulletin. 2018;**135**:915-922. DOI: 10.1016/j.marpolbul.2018.07.029

[7] Ertürk Bozkurtoğlu SN. Modeling oil spill trajectory in Bosphorus for contingency planning. Marine Pollution Bulletin. 2017;**135**:57-72. DOI: 10.1016/j.marpolbul.2017.09.029

[8] French-McCay D, Crowley D, Rowe JJ, Bock M, Robinson H, Wenning R, et al. Comparative risk assessment of spill response options for a deepwater oil well blowout: Part 1. Oil spill modeling. Marine Pollution Bulletin. 2018;**133**:1001-1015. DOI: 10.1016/j.marpolbul.2018.05.042

[9] Kindt JW. International environmental law and policy: An overview of transboundary pollution. San Diego Law Review. 1986;**23**:583-609

[10] McCarthy EM. International regulation of transboundary pollutants: The emerging challenge of ocean noise. Ocean and Coastal Law Journal. 2001;**6**:257-292

[11] Goldman R, Biton E, Brokovich E, Kark S, Levin N. Oil spill contamination probability in the southeastern Levantine basin. Marine Pollution Bulletin. 2015;**91**:347-356. DOI: 10.1016/j.marpolbul.2014.10.050

[12] Etkin DS. Analysis of US Oil Spillage. Washington DC: API Publication 356, American Petroleum Institute; 2009. 71 pp

[13] Alpert P, Neeman BU, Shay-El Y. Climatological analysis of Mediterranean cyclones using ECMWF data. Tellus. 1990;**42A**:65-117

[14] Kostopoulou E, Jones PD. Comprehensive analysis of the climate variability in the eastern Mediterranean. Part I: Map-pattern classification. International Journal of Climatology. 2007;**27**:1189-1214

[15] Pettenuzzo D, Large WG, Pinardi N. On the correction of ERA-40 surface flux products consistent with the Mediterranean heat and water budgets and the connection between basin surface total heat flux and NAO. Journal of Geophysical Research. 2010;**115**(C06022):15. DOI: 10.1029/2009JC005631

[16] Robinson AR, Golnaraghi M. The physical and dynamical oceanography of the Mediterranean Sea. In: Malanotte-Rizzoli P, Robinson AR, editors. Ocean Processes in Climate Dynamics: Global and Mediterranean Examples. Dordrecht: Kluwer Academic Publishers; 1994. pp. 255-306

[17] Malanotte-Rizzoli P, Artale V, Borzelli-Eusebi GL, Brenner S, et al. Physical forcing and physical/biochemical variability of the Mediterranean Sea: A review of unresolved issues and directions for future research. Ocean Science. 2014;**10**(3):267-279. DOI: 10.5194/os-10-281-2014

[18] Menna M, Poulain P-M, Zodiatis G, Gertman I. On the surface circulation of the Levantine sub-basin derived from Lagrangian drifters and satellite altimetry data. Deep Sea Research Part I. 2012;**65**:46-58. DOI: 10.1016/j.dsr.2012.02.008

[19] Rosentraub Z, Brenner S. Circulation over the southeastern continental shelf and slope of the Mediterranean Sea: Direct current measurements, winds, and numerical model simulations. Journal of Geophysical Research. 2007;**112**(C11001):21. DOI: 10.1029/2006JC003775

[20] Brenner S, Gertman I, Murashkovsky A. Pre-Operational Ocean forecasting in the southeastern Mediterranean: Model implementation, evaluation, and the selection of atmospheric forcing. Journal of Marine Systems. 2007;**65**:268-287. DOI: 10.1016/j.jmarsys.2005.11.018

[21] Blumberg AF, Mellor GL. A three dimensional coastal ocean circulation model. In: Heaps N, editor. Three Dimensional Coastal Ocean Models. Washington DC: American Geophysical Union; 1987. pp. 1-16

[22] Simoncelli S, Fratianni C, Pinardi N, Grandi A, Drudi M, Oddo P, and Dobricic, S. Mediterranean Sea physical reanalysis (MEDREA 1987-2015) (version 1). [Data set]. Copernicus Monitoring Environment Marine Service (CMEMS). 2014. DOI: 10.25423/medsea_reanalysis_phys_006_004

[23] Dee DP, Uppala SM, Simmons A, Berrisford P, Poli P, Kobayashi S, et al. The ERA-interim reanalysis: Configuration and performance of the data assimilation system. Quarterly Journal of the Royal Meteorological Society. 2011;**137**:553-597. DOI: 10.1002/qj.828

[24] Lardner R, Zodiatis G. An operational oil spill model in the Levantine Basin (Eastern Mediterranean Sea). International Symposium on Marine Pollution. 1998;**10**:5-9

[25] De Dominicis M, Pinardi N, Zodiatis G, Lardner R. MEDSLIK-II, a Lagrangian marine surface oil spill model for short term forecasting—Part 1: Theory. Geoscience Model Development. 2013;**6**:1851-1869. DOI: 10.5194/gms-6-1851-2013

[26] Coppini G, De Dominicis M, Zodiatis G, Lardner R, Pinardi N, Santoleri R, et al. Hindcast of oil-spill pollution during the Lebanon crisis in the eastern Mediterranean, July-August 2006. Marine Pollution Bulletin. 2011;**62**:140-153. DOI: 10.1016/j.marpolbul.2010.08.021

[27] Zodiatis G, Lardner R, Solovyov D, Panayidou X, De Dominicis M. Predictions for oil slicks detected from satellite images using MyOcean forecasting data. Ocean Science. 2012;**8**:1105-1115. DOI: 10.5194/os-8-1105-2012

[28] REMPEC. Agreement on the sub-regional systems for preparedness and response to major marine pollution incidents in the Mediterranean

[Internet]. 1995. Available from: http://www.rempec.org/admin/store/wyswigImg/file/Information%20resources/Other%20Meetings-Activities/Contingency%20planning/Subregionals%20agreements%20and%20plans/Cont%20agreement-CEI.pdf [Accessed: 13 February 2019]

[29] UNEP. Implementation agreement on the sub-regional marine oil pollution contingency plan signed by Cyprus, Greece, and Israel [Internet]. 2018. Available from: http://web.unep.org/unepmap/implementation-agreement-sub-regional-marine-oil-pollution-contingency-plan-signed-cyprus-greece-and [Accessed: 13 February 2019]